Building sub-contract management

General Editor: Colin Bassett, BSc, FCIOB, FFB

Construction Management in Practice, *R. F. Fellows, R. Newcombe, D. A. Langford and S. A. Urry*

Related titles

Construction Projects: their Financial Policy and Control, *R. A. Burgess*
A New Approach to the (DOM/1) Standard Form of Building Contract, *Glyn P. Jones*
A New Approach to the (JCT) 1980 Standard Form of Building Contract, *Glyn P. Jones*
A New Approach to the (JCT) 1980 Standard Form of Nominated Sub-Contract, *Glyn P. Jones*
Casebook of Arbitration Law, *J. Parris*
Building Contract Conditions (with 1982 Supplement), *R. Porter*
Building Contracts – a Practical Guide 4th edition, *D. F. Turner*

Building
sub-contract
management

James Franks

Construction Press
London and New York

Construction Press
an imprint of:
Longman Group Limited
Longman House, Burnt Mill, Harlow
Essex CM20 2JE, England
Associated companies throughout the world

Published in the United States of America
by Longman Inc., New York

First published 1984

British Library Cataloguing in Publication Data
Franks, James
   Building sub-contract management.
   1. Buildings—Contracts and specifications—
   Great Britain   2. Subcontracting—Great
   Britain
   I. Title
   624'.068   TH425

   ISBN 0-582-30518-7

Library of Congress Cataloging in Publication Data
Franks, James, 1927–
   Building sub-contract management.

   Bibliography: p.
   Includes index.
   1. Construction industry – Subcontracting – Management.
I. Title.
HD9715.A2F74 1984   692'.8   83-5357
ISBN 0-582-30518-7

Set in 10pt Linotron 202 Plantin
Printed in Great Britain by The Pitman Press Ltd., Bath

Contents

Foreword

Sub-contracting the constituent parts of the building process to specialist organisations can be one of the most efficient ways of carrying out building works, but much depends upon the efficiency of the specialists. These specialist sub-contractors have not in the past received a great deal of guidance or help with regard to the various matters that need to be dealt with from day to day.

Thus, there is a clear need for a book which endeavours to relate the role of the sub-contractor, as derived from the legal framework of the sub-contract, to the other disciplines and skills necessary to good management.

Of course, in so far as difficulties do arise from time to time for which the solution is not obvious, it should be remembered that while standard forms of contract or sub-contract may cater for most eventualities, they would have to be far too detailed and complex for ordinary use if it were expected that they should provide a complete answer for every forseeable problem. Such problems often derive from only being able to see one viewpoint. In such circumstances, it is worth bearing in mind the words spoken by Humpty Dumpty:

'When I use a word it means just what I choose it to mean – neither more or less.'   (Lewis Carroll: *Through the Looking Glass*)

I am sure that readers of this book will find in its pages the basis for a thorough understanding of sub-contracting in the building industry.

Roger Wakefield
September 1983

Acknowledgements

Early in 1978 I was invited by the Electrical Contractors' Association to contribute a series of articles on the subject of sub-contract management for publication in *The Electrical Contractor* which aroused considerable interest and led to requests to talk to sub-contractors on contract matters. The articles provided the starting point for this book.

To Brian Thornton, director of Stewart Wrightson (Construction Insurance Services) for assistance with the chapter on *Insurance*, to my colleague David Coles for assistance with surveying and estimating matters, to John Tuck for use of his claims material, to Peter Roper, editor of *Building Trades Journal* for allowing me to use material from some of my articles and other extracts from that journal, to Terry Murphy for access to his collection of managers' proverbs some of which were too colourful for inclusion in these pages and to Tony Rea, editor of *Electrical Contractor* I am very grateful and I offer my thanks.

So that points of contract and details of procedures may be discussed in a practical manner and to best effect this book uses as its model the Joint Contracts Tribunal's *Standard Form of Building Contract* (1980 Edition) and the Nominated Sub-Contract documents. Quotations from the documents are apparent in the text and the author gratefully acknowledges the kind permission of the publisher, RIBA Publications Ltd, to reproduce these extracts from the JCT forms, and also the kind permission of the National Federation of Building Trades Employers to reproduce extracts from NSC/4.

But most of all my thanks to my wife, Maureen, without whose patience, research and assistance with 'assembly' the book would never have materialized.

James Franks
Fitzroy House,
10 High Street,
Lewes, BN 2AD

Abbreviations used in the book

CIOB	Chartered Institute of Building
COD	Concise Oxford Dictionary
DOE	Department of the Environment
DOM/1	JCT Domestic Sub-Contract
ECA	Electrical Contractors Association
FASS	Federation of Associations of Specialist Engineering Contractors
HVCA	Heating and Ventilation Contractors Association
IPS	Incentive Payment Scheme
JCT	Joint Contracts Tribunal
NCC	National Consultative Council
NEDO	National Economic Development Office
NFBTE	National Federation of Building Trades Employers
NJCC	National Joint Consultative Committee
NSC/1	JCT Standard Form of Nominated Sub-Contract Tender and Agreement, 1980 Edition
NSC/2	JCT Standard Form of Employer/Nominated Sub-Contractor Agreement, 1980 Edition
NSC/3	JCT Standard Form of Nomination of Sub-Contractor, 1980 Edition
NSC/4	JCT Nominated Sub-Contract, 1980
OED	Oxford English Dictionary
PD–OR	Paul Dickson, *The Official Rules*, Arrow Books, 1981
RICS	Royal Institution of Chartered Surveyors
SF 80	Standard Form of Building Contract, 1980 Edition
SMM	Standard Method of Measurement of Building Works

Introduction

The primary aim of this book is to provide a practical management manual for the many specialist contractors who are engaged as sub-contractors by main contractors on building projects. These specialists already carry out some 80 per cent of the actual building work and they have an increasingly important role. The administration and management of sub-contract works present different problems from those of the general contractor yet very few books are devoted especially to these specialist sub-contractors' needs.

Generally speaking the arrangement of the book follows the course of a sub-contractor's 'contract' or 'job', starting with 'obtaining work' and ending with 'the final account'. The scope of the chapters may be seen in the Contents.

The secondary aim is to meet the needs of the students who plan to make careers in specialist firms; many in various forms of engineering services which are essential to the proper operation of modern buildings. With their needs in mind an attempt has been made to include some of the 'principles' and theoretical background to the subject which the experienced contractor may find superfluous. It should not be difficult for the reader to select the aspects which are of greatest interest.

Finally, as far as readership is concerned (and it is appreciated that no book of this type can be all things to all men) it is hoped that at least part of the book will be of interest to the consultant engineers and their students whose practices extend into the realm of building projects. The role of these consultants is becoming increasingly important and their involvement in building projects is in many respects similar to that of the specialist contractor whose work they design and specify. Experience gained teaching on degree and diploma courses for structural and building services engineers indicates that many of those courses' syllabi require quite extensive knowledge of contract administration and other matters dealt with in this book. No attempt has been made to discuss all contract administration matters but the principal aspects have been included.

The specialist contractor's
PURPOSE
is to make a profit by providing a sub-system
which meets the client's needs. The
TASKS
which he has to undertake to serve the purpose are:

identify his market and forecast demand
establish his aims and objectives
obtain work
plan, organize and control, and
obtain payment for his work

To carry out his tasks requires
SKILLS
which include:

taking decisions
managing people
delegating
communicating, and
managing his own time

The specialist contractor has
RESPONSIBILITIES
to his organization, himself and to those with whom he comes in contact

Chapter 1

The construction industry as a market for the specialist (sub-)contractor

No business will survive unless there is a market for the products or services which the business provides. Enthusiastic, would-be entrepreneurs with more optimism than experience frequently ignore that basic law of supply and demand.

What prospects does the construction industry offer for the specialist contractor?

The role of the construction industry

Construction industries, universally, are remarkably similar in that they are major industries made up of numerous small firms. The construction industry in the USA, for example, accounts for 11 per cent of the gross national product and is bigger than the auto or steel industries but much more diversified, and it comprises a great number of small businessmen. There are 800,000 firms of which only a third maintain a permanent payroll. Most have only one to four employees.

The construction industry in France has some 33,000 firms, 28,000 of which have only a 'handful of workers' (Westminster 1972).

The British construction industry accounts for almost 10 per cent of Britain's gross national product. This is some 4 or 5 per cent less than the industries of France and West Germany but it represents a substantial market for British specialist contractors (House 1977).

Apart from its size, the British construction industry has other attractions for the entrepreneur. Like the industries of the USA and France, referred to above, it has some 73,500 firms, of which more than 50,000 employ fewer than eight people. In addition, there are approximately 200,000 self-employed workers. The importance of construction industries is frequently the reason for them being used as 'regulators of the national economy' by their respective governments.

The State is usually a major client of it's construction industry (more than 50 per cent of most countries' construction work is directly or indirectly for the State) and this makes it comparatively simple for a government to stimulate or reduce growth in order to regulate its national economy.

It has been suggested, too, that a government will occasionally stimulate

1

demand for political, as well as for economic, reasons. By creating work in the year or so before an election it is possible for a government to give electors an impression of well-being which will encourage them to re-elect the government for a further term. The large size of the construction industry and the effect which a buoyant construction industry has on the national economy as a whole makes construction an obvious choice for stimulation which could (and occasionally does) create a trade cycle geared to elections.

The extent to which such manipulation of the British construction industry is responsible for the peaks and troughs to which the industry is prone is by no means certain but there is no doubt that the building work-load fluctuates greatly. A number of factors probably contribute to the fluctuations. 'The general conclusion [is] that individual firms of contractors will have been quite substantially affected by the fluctuations due to stop and go, although the effect over the industry as a whole was less marked' (Hillebrandt 1979)

Specialist contractors engaged in the construction industry are, then, vulnerable as the second half of the 1970s, in particular, demonstrated all too dramatically.

The building process

A prerequisite of the building process is a client with a need for a building. The need is usually for a 'shelter' to perform a specific function or for a service in connection with a shelter.

The shelter is usually a 'unique product' which is constructed in an 'open-air factory' – the building site – by an ad hoc 'team' comprising designers, builders, and specialist contractors who have frequently not worked together before and who will probably not do so again. Frequently this team is working for a client who is building for the first time and who does not really know what he wants.

Assuming the foregoing statements are correct, how does this unhappy 'state of the art' arise?

The principal problem is that the client's needs for his shelter are often quite sophisticated and the client lacks the knowledge required to *state* his needs and to *communicate* them to the persons who will be doing the actual building. In this event he must employ a person with the requisite skills to translate the client's need into a design and communicate the design to the builders. In practice the complexity of modern building is such that it is frequently beyond the ability of one person to undertake the whole design and many persons are required to build. Good communication between the parties is obviously highly desirable.

The actual building is a matter of bringing together the necessary resources at the right place and at the right time. The resources; the labour, plant, and materials must be of appropriate quality and quantity to meet the design which is normally expressed in the form of drawings and specifications. The organization and control of the resources requires considerable managerial skills.

Very few clients are able to build without regard to cost and it is, therefore, necessary to have a means of cost control built into the construction process. How then, does the process operate?

The organization structure which has been established in Britain is similar in most, but not all, respects to the structure used in most other countries. It emerged in the latter part of the nineteenth century and it has remained virtually unchanged until the present time, although in recent years considerable doubts have been expressed about its suitability for all types of project. The relationships and the lines of communication involved in the organization structure were the subject of a survey commissioned by the Tavistock Institute *Communication in the Building Industry* in 1965. The authors of the Tavistock Report, (Higgin and Jessop) expressed the relationships in the building 'team' in the diagram shown in Fig. 1.1

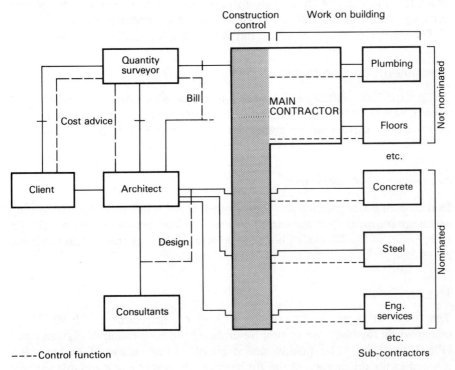

Fig. 1.1 Relationship between parties to building contracts (after Higgin and Jessop 1965)

The parties to the contract

The roles of many of the parties to the contract are discussed in detail in later chapters but in order to indicate their part in general terms they may be outlined as follows.

The Client (Employer)

'Employer' is the term used for the client in the Standard Form of Building Contract. In the wording used in the Articles of Agreement to SF 80 it is the employer who is desirous of having the 'Works' built. He undertakes to pay the contactor the contract sum. The employer also employs the architect.

The Contractor

The Contractor undertakes to carry out and complete the Works shown upon the Contract Drawings and described by or referred to in the Contract Bills (of Quantities) and in the Articles of Agreement SF 80, cl. 2.1). The contractor is referred to as the 'Main Contracter' in the Nominated Sub-Contract, NSC/4, which he enters into with the nominated sub-contractors. The contractor may enter into contracts with sub-contractors who are *not* nominated.

The contractor is occasionally referred to as the 'general' contractor.

Nominated Sub-Contractor

'Nominated Sub-Contractor' is defined in clause 35 of Part 2 of SF 80. The 'nomination' is made by the architect with agreement of the contractor. The work which may be carried out by nominated sub-contractors is in no way restricted to the type of work shown opposite nominated sub-contractors in Fig. 1.1.

Domestic Sub-Contractors

Domestic Sub-Contractors are those, other than nominated sub-contractors, which are shown as 'Not nominated' in Fig. 1.1. The contractor may not sub-let any portion of the Works without the written consent of the architect (SF 80, cl. 19.2).

The Architect

The Architect is referred to in the Articles of Agreement to SF 80. He is employed by the employer to prepare or direct the preparation of drawings and bills of quantities. 'The quantity and standards of materials or of workmanship is a matter for the opinion of the Architect, such quality and standards shall be to the reasonable satisfaction of the Architect' (SF 80, cl. 2.1).

The architect is empowered to issue instructions to the contractor (SF 80, cl. 4), to vary the Works (SF 80, cl. 13), to certify payments and to give extensions of time, etc. (SF 80, cl. 25).

The Quantity Surveyor

The Quantity Surveyor is referred to in Article 4 of the Articles of Agreement of SF 80. He prepares the bills of quantities which, when they have been priced by the contractor, become the *Contract Bills*. He values variations (SF 80, cl. 13) and he may be instructed to ascertain the amount of loss and/or expense caused by matters materially affecting the regular progress of the works (SF 80, cl. 26). He makes recommendations to the architect regarding the sums which should be paid to the contractor and nominated sub-contractors by way of payments on account (SF 80, cl. 30).

The quantity surveyor was described in the Banwell Report (Banwell 1964) as the economist of the Construction Industry; this is an extension of his role as the client's advisor, cost planner and controller.

The Consultants

The Consultants shown in Fig. 1.1 are not mentioned in the Standard Form of Building Contract but they include the structural engineer, building services engineers and other experts in specialist matters associated with building projects.

Separation of design from construction

Higgin and Jessops's diagram illustrates what Sir Harold Emmerson reported when he was asked by the then Minister of Works to make a quick review of the problems facing the construction industries in 1962, namely, that 'in no other important industry is the responsibility for design so far removed from the responsibility for production' (Emmerson 1962). The divorce of design from production is at the root of most of the communication problems in the building industry and the sub-contractor probably suffers more than most parties because he is at the end of the line.

A simple example demonstrates the communication problem. Suppose the client on a project for a factory in course of construction discovers that new regulations will make it necessary for him to provide higher standards of air-conditioning in part of the factory than those originally specified. If we look at the diagram we see that he must contact the architect who will consult the engineer about types of plant which will provide the appropriate standard of air-conditioning. The consultant engineer will advise the architect of the alternative types of equipment which will do the job and he will probably make recommendations about the solutions. The architect will select the equipment which he considers fits in with the design of the building and ask the consultant engineer to provide drawings and specifications for the proposed equipment.

If the equipment requires larger-section ducting it may be necessary for the structural engineer, a different man from the consultant engineer responsible for the engineering services, to redesign the structural member through which ducting has to pass. (It is a fact of building life that any alteration during the course of building works seems almost always to involve the part of the structure which holds the whole building together.)

The architect will, therefore, pass the consultant engineer's drawings to the structural engineer and obtain revised structural drawings from him. In any event, when the consultant engineer and the structural engineer have produced their technical solutions the architect has to redesign the wall panelling, suspended ceiling or whichever part of the room finishings will accommodate the new equipment.

The architect then issues revised drawings and instructs the main contractor, who instructs the sub-contractor who places an order for the air-conditioning equipment with a supplier and so on . . .

As if that complicated line of communications was not enough in itself Sir Harold Emmerson found when making his review of problems that 'there is all too often a lack of confidence between architect and builder amounting at its worst to distrust and mutual recrimination' and 'even at their best, relations are affected by an aloofness which cannot make for efficiency . . .'

To say the least, the outlook for the parties to building contracts is not encouraging.

But fortunes are made in the British construction industry by able and ambitious men even if the bankruptcy rate is higher than almost any other industry. It is a highly competitive industry with small profit margins. Nevertheless, the size of the industry and the fact that it may be entered with virtually no capital because the client usually makes frequent 'interim' payments on account make the building industry an attractive market for the entrepreneur.

References and bibliography

Banwell committee (1964) *The Placing and Management of Contracts for Building and Civil Engineering work*. HMSO.

Emmerson, Sir H. (1962) *A Survey of Problems Before the Construction Industries*. HMSO.

Higgin, G. and Jessop, N. (1965) *Communication in the Building Industry*. Tavistock Institute, London.

Hillebrandt, P. M. (1979) *Economic Theory and the Construction Industry*. Macmillan Press, London.

House Information Services (1977) *Construction Industry, Europe 1976–77*.

Westminster Chamber of Commerce Building Group Mission to Europe (1972) *Building in the EEC*. Westminster Chamber of Commerce, London.

Chapter 2

Aims, objectives, corporate plans and policy

'If you're not in business for fun or profit,
what the hell are you doing here?', (Townsend 1970).

'There is a well-established theory of man as a "wanting" animal' (Maslow 1970).

Man is regarded as an animal who is never satisfied because as one want is appeased he invents another to take its place – and so on. Women are, of course, no different in this respect.

Hierarchy of wants

The situation which best illustrates this aspect of man's character is a shipwreck. He is first seen swimming for his life searching for a raft, lifeboat or piece of flotsam – he wants to survive. He finds his raft but as soon as he has recovered his breath he wants water to drink, food to eat and shelter from the wind, or rain, cold or heat depending on where he was shipwrecked. (Fortunately for specialist contractors on building projects, shelter is high in the hierarchy of man's wants.)

His hunger and thirst satisfied and his physical comfort being assured, he looks for another want. If there are fellow survivors on the raft he wants to establish himself in their eyes. He needs status. And so it goes on.

The initial wants – survival, food, drink and shelter – are common to all men but beyond these each has his own *hierarchy of wants*. These vary from person to person. One man's video recorder is another man's dinner for two at the Savoy.

The root of Maslow's argument is that the more one want is being satisfied, the less the satisfaction matters.

Maslow's work is concerned primarily with human motivation but there is a relevance to aims, objectives and corporate plans in his theories.

A specialist contractor starting up in business is much more concerned with obtaining sufficient work to feed and house his family, to stay in business to obtain sufficient cash to pay his bills than he is with making long-term plans.

As an anonymous manager puts it so succinctly, 'When you're up to your arse in alligators it's difficult to remember the original objective was to drain the swamp.'

Nevertheless, as the specialist contractor moves away from the survival items in his hierarchy of wants he is able to set objectives for his firm and attempt to plan for the future rather than react to events. For the smaller, more recently

7

founded firm *management by objectives* provides a method of corporate planning which is generally appropriate for its needs.

Management by objectives

Management by objectives has been described as 'a dynamic system which seeks to integrate the company's need to clarify and achieve its profit and growth goals with the manager's need to contribute and develop himself' (Humble 1974).

The key to management by objectives is continuity – there is a continuous, step-by-step process of review and measurement of progress. The specialist contractor cannot simply make his plan for the next five years and put it in a drawer. There must be regular and frequent reviews of progress during which actual achievement is measured against the target which the specialist contractor had set himself for the previous period.

Following on a critical review the specialist contractor states (or re-states) the firm's strategic and tactical plans. *Strategic* plans are those which 'serve the ends' – the broader, longer-term plans which will take the firm where it hopes ultimately to be. *Tactical* plans are the day-to-day plans. Both terms are borrowed from military vocabularies where strategy tends to be the province of generals and tactics are concerned with 'actual contact with the enemy'.

The second step is *'definition'*. If the specialist contractor is, say, a partnership, each partner defines the key results and performance that he must achieve, in line with company objectives. He commits himself to achieving these objectives.

In order that the firm shall achieve its objectives it is usually necessary for each partner to have the right conditions in which to maintain or (and preferably) improve his performance. These conditions, which may include setting up an appropriate organization structure, must be provided by the firm as part of its package plan.

The partners must decide on a system for providing control information which will enable them to monitor their achievement and make better and quicker decisions. They must have plans for management and technical training for themselves and for personnel who will provide the firm's management for the future.

The partners should be allowed as much autonomy, freedom and flexibility as possible. The controls should be simple, economical and relevant. The training and development should have positive aims rather than be indulged in because it is fashionable.

Management by crisis

Management by objectives is a basic approach to corporate planning which is suitable for the smaller specialist contractor. It is often cited as an alternative to 'management by crisis' where the manager fights whichever fire is the most threatening at a given time. Management by crisis is discussed in Chapter 7.

Strategic and tactical plans

Corporate Planning

Corporate planning is often considered to be something for large organizations.

This is no doubt because 'corporate' suggests corporations and we tend to think of corporations as very large organizations. While it is true that corporate planning started in the large organizations and is largely practised by them, small organizations also need a corporate plan if they are to succeed.

They need aims, they need objectives, and plans to turn their aims and objectives into realities. Studies of firms show that a surprising number of them – not just the large firms – make corporate plans. That is not to say that their methods of corporate planning are highly sophisticated; the fact that they make a conscious commitment to planning their future is the important factor.

Corporate planning is not a new concept but like so many subjects in management thinking it came into fashion in the 1960s.

Hussey's definition (1978) of corporate planning is:

'A comprehensive, continuous process of management looking towards the future; which is responsive to change in the external environment. It is concerned with both strategic and operational plans, and through participation develops plans at the appropriate levels within the organization. It includes methods of monitoring and control, and is concerned with both the short and the long term'.

Let us look into the constituents of that definition.

'A comprehensive, continuous process'

The plan should cover all aspects of the organization's activity. Eight 'key areas' of activity may be identified: marketing, innovation, human organization, financial resources, physical resources, productivity, social responsibility and profit requirements (Drucker 1974). The first in the list in all respects is marketing.

The marketing objective

Specialist contractors whose businesses are rooted in the building industry often believe that marketing does not concern them. They suggest that they do not make or sell a product in the same way as, say, detergent or pet-food manufacturers. It is true that their products are different from those of the manufacturers mentioned above but they nevertheless operate in a market-place and need a plan for the future.

The specialist contractor who restricts his activities to purely contract work provides a service as much as a product. There was a time when the competitive tender was the only source of work for the specialist contractor. The competitive tender is by no means dead but it is much less important than it was as a means of selecting the contractor for the project.

Services tend to sell by reputation, by quality, rather than by price. This incidentally, has been a major plea of the professional institutions when putting their cases for the retention of scale fees to the Monopolies Commission.

The last few years have shown a marked increase on the part of service industries and the professions to actively market their services. In the building industry, in particular, one of the most remarkable changes has been the way in which providing service and the ability to market a service has become increasingly more important than the most competitive price.

More and more specialist contractors have a second (or more) string to their bow. Timber treatment firms have diversified into damp-proof course installation, aluminium component companies have entered the secondary window installation business, etc.

Clearly, this specialization needs a corporate marketing plan – the market is at the centre of any business enterprise; without a market for its goods a business cannot exist.

Several courses are open to the corporate planner. When setting marketing objectives he can:

(a) set targets to improve the performance of existing services or products in the existing markets (e.g. increase productivity on his current projects, reduce contract periods, etc.);

(b) find new markets for the existing services or products (e.g. seek inclusion on new clients' tender lists);

(c) develop new services or products for the existing market – for clients with whom he has developed a good relationship during contracts he has obtained by competitive tender;

(d) develop new services or products for new markets (e.g. patent damp-proof treatment for insertion in old buildings).

The objectives set the strategy for the future policy of the company. The policy should be stated as briefly as possible. Indeed, if it cannot be stated briefly it is probably not 'policy'. Policy is sometimes described as the 'concentration decision'.

An example of a concentration decision is provided by Barratt Developments, the firm of specialist housebuilders, whose policy is to build houses selling at two-and-a-half years of the purchaser's income – an index-linked policy statement!

A firm uses its objectives as the strategy to achieve its policy. Business experts, talking or writing about business in general, usually point to the need for the firms to decide in which part of the market, with which service/s and/or product/s it wishes to be the market leader. Much is made of the importance of having a major share of the market. It is pointed out that the 'marginal' supplier is vulnerable in times of economic setback.

It is easy to dismiss such comments as not applicable to building firms. Building firms, we are fond of saying, are different. Are they? When thinking through its corporate plan the specialist sub-contractor should carefully consider if the rules applicable to other industries apply also to their own, before the important decisions are made. There must be a reason for the building industry having the highest bankruptcy rate.

Innovation objectives

Marketing is concerned with the market 'as it is'! If business is not market orientated, is not what the client wants, we must determine what our business should be. The 'customer is right' and we must change, innovate, our business to be what it should be.

It is possible to bring about innovation in three ways. First, we can innovate our service or product. Developing a design-and-build service or a new building form for, say, light industrial use, are examples of such innovation.

Second, there is innovation in the market or client needs. The collapse of the UK building market in the mid-1970s following the 1973 oil crisis provides an example of a changing market.

British specialist contractors found little work available at home and they were forced to look to new markets overseas.

The turnover of most contractors dropped considerably and some went out of business as a result of their failure to respond to the changes in the market. The drastic oil crisis is an extreme example but changes in clients' needs have been a regular feature of the post-war building scene.

Clients' needs seem to go in phases. After the war housing was followed by industrial buildings and schools (to meet the population explosion caused when people picked up their lives when the war ended), followed by offices and hospitals, followed by rehabilitation of existing property and so on. As soon as a contractor has become experienced in a particular type or work, that type has been superseded by another.

The third type of innovation concerns the activities and skills required to perform the services and make the products and market them.

Innovating activities and skills follows the other kinds of innovation. Whilst the traditional building specialist contractors are still alive, in spite of the apparent lack of concern of the building industry's problems by successive governments, they have to a considerable extent changed to meet new demands.

Dry linings and dry partitions, for example, have led to the development of new skills and specialists have emerged to handle these and other new products which have appeared on the market from time to time.

The innovation objective is, then, a conscious attempt to anticipate future client needs and markets and requires an ability and/or willingness to make forecasts. It requires a positive commitment by the contractor to forecast where the next, or even next but one, market growth area will be and to specify and quantify it.

It is often suggested that smaller organizations are better placed than large to have their finger on the market pulse and that they are willing and able to innovate more quickly. Experience has certainly shown that specialist contractors, regardless of size, who are able to innovate have survived and even grown when others less responsive have gone to the wall.

The resources for the plan

Human organization, financial resources and physical resources are three of the key areas of the organizations activity identified above. Successful production, whatever its end-product, requires an ability to bring together all the necessary resources: manpower, plant materials and management (the last is the catalyst) at the right time, in the right quantities and in the most economical manner in order to meet the client's needs.

Objectives are required to ensure an adequate supply and utilization of these resources in order to achieve optimum productivity. As far as manpower is concerned, management must set out a programme of recruitment and training for the future.

Larger firms will almost certainly require the services of a training officer and management should include him in its decision-making processes. It is appreciated that not all contractors are sufficiently large to warrant a personnel and training officer of their own and such firms may find their trade association or training board useful alternatives.

Financial planning and control is discussed in Chapter 7, sufficient for the time being to say that a corporate plan which does not include a budget for accomplishment of the plan is not worth the paper it is written on. Conversely the financial planner who does not fully appreciate the significance of the corporate plan in which his budget will be used cannot make adequate plans.

The specialist contractor who manufactures any of the components which he uses in his day-to-day business, and many do this, must have a plan for renewing or improving his plant, and stocking up against the occasional shortages of materials to which the building industry seems to be so prone.

To summarize the position as far as resources are concerned the specialist contractor should ask himself two questions. First 'these are the resources we need; how can we go about obtaining them?'

Second, and perhaps more important: 'these are the resources which are available; what must we do to use them to our best advantage?' The decisions he makes on the basis of the answers he gives himself to these questions may determine his future success or failure.

Productivity objectives
Management needs a means of measuring its attainment and productivity provides the most direct and positive means. For individual projects, budgets (targets, yardsticks or standards – all names for productivity measures) should be set up using the estimate as their basis.

Because building projects take a considerable time to complete it is important that the total project is broken down into a number of operations which can be measured and controlled at regular and frequent intervals.

Money and time are the ultimate units of measure. The manpower, plant and materials which are the ingredients of the operations can all be measured and expressed in money and time terms. Indeed, when contractors build up 'unit rates' in order to prepare their tender they often use 'constants' which represent notional amounts of manpower, plant and materials.

It makes good sense to use these unit rates to measure productivity and at the same time profitability. Unit rates are, however, not always suitable as productivity measures. The units are often too small. Contractors' programmes are normally operation based and it follows from this that operations usually provide a better basis for productivity measurement than units.

Operations can, then, be used as cost centres and as project productivity objectives.

Methods by which these productivity objectives may be monitored and by which control may be exercised are discussed in Chapter 7.

Social responsibility objectives

Social responsibility extends to ensuring that contractors cause the least disturbance possible during the course of their works. The company's policy statement can do much in this connection by setting standards for noise reduction, removal of silt in containers and not by discharge into public drains, etc., etc. In addition to their contractual obligations specialist contractors should feel that they have a social responsibility to the main contractor and to the other sub-contractors with whom they are working.

The profit objective

'If you are not in business for fun or profit what the hell are you doing here?' Robert Townsend (1970) asks that question on the dust cover of his amusing but perceptive book about the management of commercial organizations entitled *Up the Organisation*.

At first reading it appears to be a rather flippant and almost irrelevant question but the more one considers it the more one realizes that it sums up the whole of one's business philosophy. What are our reasons for being in business? Are we in business for fun or profit?

A surprising number of contractors on building projects are in business for fun rather than for profit. In saying that one assumes that one interprets 'fun' as meaning because one enjoys the business. Over a period of years one meets many who years ago made quite enough money on which to retire but who keep going because they actually enjoy being part of a team creating, what is frequently, a unique building.

But whatever his motives for working and however much fun his business may be the specialist contractor still needs profit for several reasons, not least, in order to survive. Profit is also needed to pay for the attainment of the company's objectives.

The specialist contractor needs a profit to cover:
(a)   risk (staying in business);
(b)   tomorrow's capital (to finance future work);
(c)   growth and innovation.

Obviously, when planning profit there is a need to ensure that provision is made for taxation. Profit is not a dirty word; it is the economic surplus which pays for social benefits and security.

The objectives the organization is considering setting have to be weighed and priorities given to them. Risks have to be taken; but business is risk-taking.

Decisions have to be made which will affect the future and budgets have to be set taking into account the availability of resources and creditability of the organization. Setting the budgets is a top management decision.

While the mechanics of budgeting is largely an accounting matter the decisions are entrepreneurial.

Preparing a corporate plan

Numerous approaches to corporate planning have been devised. Of these, four are discussed as being appropriate for specialist contractors engaged on building projects.

It should be kept in mind that no two firms have identical needs and that it is not therefore possible to lay down a rigid method or methods. The individuals responsible for undertaking the corporate plan must themselves decide on the approach to adopt, which will vary from firm to firm.

Stocktake, plan, review approach

This approach has probably been in use longer than most and is considered by some management philosophers to be rather out-of-date. It is, nevertheless, an approach adopted by many firms. Management thinking is second only to women's dress fashions in its tendency to adopt and discard fashions. It would be a brave man who would denigrate an approach which appears to be working satisfactorily.

The stocktake part of the stocktake, plan, review approach consists of making a systematic record of all aspects of the company to find out just what is happening.

The examination would usually start with a study of the organization and management of the firm looking for its strengths and weaknesses. The flow diagram in Fig. 2.1 indicates an approach which has been used to good effect.

Having determined the company's strengths and weaknesses it is possible to set objectives for the future along lines discussed above.

Setting objectives is one thing but achieving them is something quite different and requires a strategy. The strategy determines the future organization structure of the company (not vice versa) and leads to the preparation of operating plans which will probably require management development, clarification of personnel roles and functions and setting standards against which performance can be measured.

Appraisal and review of performance is possible if key areas are made cost centres with pre-determined budgets, the preparation of which is an important aspect of the management of money.

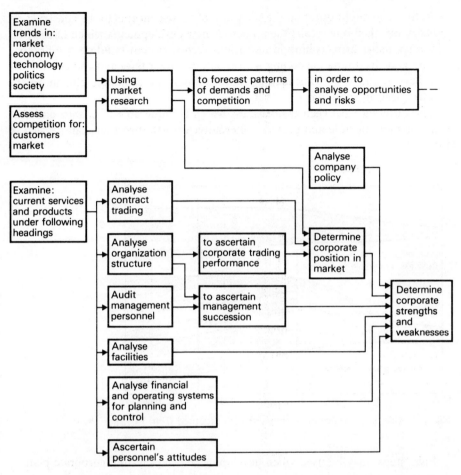

Fig. 2.1 Stocktaking to determine company strengths and weaknesses (after Lumsden 1971)

Step-by-step approach

The step-by-step approach concentrates on markets with a record of innovation and the development of new business. It aims to exploit the company's skills and resources in the best market environment. It is essentially an approach which provides flexibility (Buckner 1974).

The primary objectives of the step-by-step approach are:

(a) setting targets for the use and development of corporate skills and resources independently of the present products;

(b) using the company's skills and resources in the best market environment;

(c) planning for flexibility by concentrating on areas having a record of innovation and development of new businesses;

(d) Selecting a business where it is possible to maintain a unique position.

At first one might question the relevance of these concepts to the construction industry but then one recalls the numerous new developments which have given rise to specialist firms (piling in its various forms, system building, proprietory roofing and dry-lining/partioning – to mention only a few) and the contractors who have found rewarding markets abroad when the economic climate in the UK has been in recession.

The difference between the business which is vulnerable to change and that which is more flexible and provides alternative growth directions can be seen in Fig. 2.2.

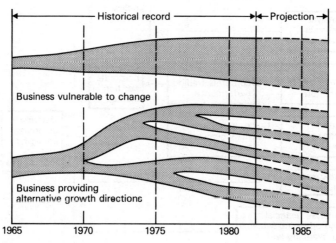

Fig. 2.2 Alternative types of business growth (adapted from Buckner 1974)

Five 'steps' may be used when developing the step-by-step corporate plan.

1. To determine the size of the long-range problem which involves:
   (a) setting quantitative objectives;
   (b) setting qualitative objectives;
   (c) projecting the present business;
2. To improve the performance of existing services or products in the present market;
3. To find new markets for present products – or new products for present markets;
4. To develop or acquire new services or products for sale in new markets;
5. To implement new product plans.

The closed-loop process has developed from the size-up to determination of objectives and so to programme preparation method, through the open-loop process to arrive at what is probably as good an approach as any for smaller companies (Gilmore 1971).

Closed-loop approach

The closed-loop approach can be considered as six progressive tasks:

1. Record current strategy;
2. Identify problems;
3. Discover the core elements;
4. Formulate alternatives;
5. Evaluate alternatives;
6. Choose the new strategy.

This simple approach, similar in many respects to the method study process with which many managers are familiar, relies for its success on well-guided meetings chaired by the managing director.

Group problem-solving sessions are arranged during which, more or less, structured questions are considered by the management team regarding each of the six tasks.

The tasks can be considered as milestones. Each milestone has to be passed, in turn, in the progress towards a solution.

A list of questions which can be asked is provided in Table 2.1. It is important to appreciate that the six tasks are not carried out on a once-and-for-ever basis. The closed-loop approach is, as its name suggests, a continuous process with provision for sizing-up the situation, formulation of strategy, organization for implementation of the strategy, controlling performance and reappraisal of current strategy.

A flow diagram showing the tasks to be undertaken in the closed-loop system is given in Fig.2.3.

The approach was developed during a research project at Cornell University (Gilmore 1971).

The process is responsive to changes, trends and threats. The planner formulates a revised economic mission, competitive approach, programme of action, etc. to suit the changing circumstances. In this context the term 'economic mission' means the kind of business the company should be in and what its performance objectives should be.

The competitive approach is concerned with finding the product-market-sales approach that will accomplish the economic mission and with deriving appropriate goals in the various areas of the company's business. The programme of action involves searching for efficient means of implementing the competitive approach.

The process illustrated in Fig. 2.3 is continuous. Current strategy is reviewed from time to time in the light of operating results, economic trends, competitors' actions and technological developments.

When business opportunities or threats appear, management asks questions designed to determine the ways and time in which strategy should be changed. If a new economic mission is called for, the planners proceed to formulate a

Table 2.1 Questions to ask when following the closed-loop process

1. *Record current strategy*:
   (a) what is the current strategy?;
   (b) what kind of business does management want to operate (considering such management values as desired return on investment, growth rate, share of market, stability, flexibility, character of the business, and climate)?;
   (c) what kind of business does management feel it ought to operate (considering management's concepts of social responsibility and obligations to stockholders, employees, community, competitors, customers, suppliers, government, and the like)?

2. *Identify problems with the current strategy*:
   (a) are trends discernible in the environment that may become threats and/or missed opportunities if the current strategy is continued?;
   (b) is the company having difficulty implementing the current strategy?;
   (c) is the attempt to carry out the current strategy disclosing significant weaknesses and/or unutilized strengths in the company?;
   (d) are there other concerns with respect to the validity of the current strategy?;
   (e) is the current strategy no longer valid?

3. *Discover the core of the strategy problem*:
   (a) does the current strategy require greater competence and/or resources than the company possesses?;
   (b) does it fail to exploit adequately the company's distinctive competence?;
   (c) does it lack sufficient competitive advantage?;
   (d) will it fail to exploit opportunities and/or meet threats in the environment, now or in the future?;
   (e) are the various elements of the strategy internally inconsistent?;
   (f) are there other considerations with respect to the core of the strategy problems?;
   (g) what, then, is the real core of the strategy problem?

4. *Formulate alternative new strategies*:
   (a) what possible alternatives exist for solving the strategy problem?;
   (b) to what extent do the company's competence and resources limit the number of alternatives that should be considered?;
   (c) to what extent does management's preferences limit the alternatives?;
   (d) to what extent does management's sense of social responsibility limit the alternatives?;
   (e) what strategic alternatives are acceptable?

5. *Evaluate alternative new strategies*:
   (a) which alternative best solves the strategy problem?;
   (b) which alternative offers the best match with the company's competence and resources?;
   (c) which alternative offers the greatest competitive advantage?;
   (d) which alternative best satisfies management's preferences?;
   (e) which alternative best meets management's sense of social responsibility?;
   (f) which alternative minimizes the creation of new problems?

6. *Choose a new strategy*:
   (a) what is the relative significance of each of the preceding considerations?;
   (b) what should the new strategy be?

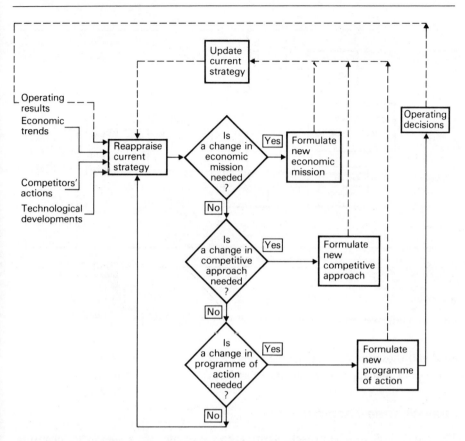

Fig. 2.3 Strategy formulation process (after Gilmore, 1971)

revised approach. This will, in all probability, require a fresh competitive approach and programme of action.

If the economic mission is considered to be sound, the competitive approach is looked at and so on until the area which needs revision is identified and a new strategy is formulated. Should the overall current strategy appear sound no revisions would be necessary and the current strategy would continue until the next review.

The broken lines in Fig. 2.3 indicates feed-back of operating results and the updating of current strategy as a result of current or revised formulation.

An essential element of the strategic review approach is, as its title suggests, its emphasis on strategy. Experience has shown that in many approaches to corporate planning the tendency is for goals and assumptions to go down the organization and for detailed plans to go up and that this approach is not necessarily the best.

An important reason for the inadequacy of this approach is that by the time the plan emerges, a great deal of work has been done on it, opinions have hard-

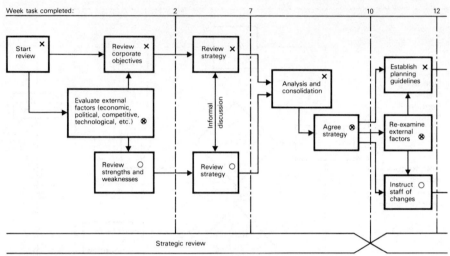

Week task completed: 2   7   10   12

Strategic review

X board or management group tasks
O departmental or divisional tasks
⊗ combined tasks

Fig. 2.4 Outline planning process (adapted from Young and Hussey 1978)

ened and those involved in the planning have become so committed to their part of the plan that they are reluctant to accept alternative proposals.

Strategic review approach

The strategic review approach attempts to overcome this problem by ensuring that the right type of analysis is carried out; that there is participative and objective discussion between the corporate top management and heads of business units; and that strategic issues are identified from both corporate and divisional perspectives.

This makes it possible for corporate plans to be thought about at the 'input' stage. Detailed planning tasks (not least being the preparation of the budget), are not attempted until the strategy has been discussed and agreed by all concerned in preparing and implementing it.

The key is, then, early consultation with and continuous participation by all levels of management.

Figure 2.4 shows an outline planning process which can be used when adopting the strategic review approach to corporate planning. It should be self-explanatory but it is worth remembering that the time-scale is only indicative and would in fact depend on factors such as whether the review was being carried out by outside consultants or by a team recruited from within the company (which could spend only limited time on the plan), the size of the organization, etc., etc.

20

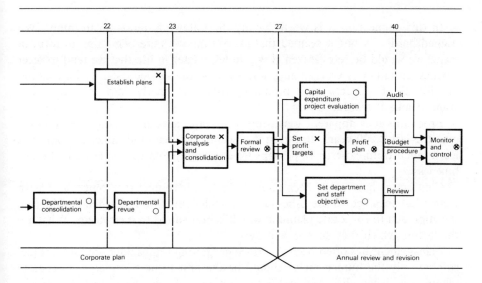

Applying corporate planning methods

It was suggested above that management by objectives would provide a suitable corporate planning approach for the smaller, more recently-founded firm. For the larger, established, firm one of the more formal corporate planning methods should be considered. Which, then, would be appropriate for a specialist contractor, founded some thirty years ago with one of the founder members as managing director holding 40 per cent of the shares, 30 per cent being held by members of his family and the balance being held by the other three directors?

How can the run-of-the-mill firm make corporate plans? Is the political/economic environment in which the firm lives so unpredictable that corporate planning is pointless?

Reference was made to the fact that a considerable number of contractors already make corporate plans so, to answer the last question first, some firms, at least, consider that they have some influence over, if not full control of, their own destinies.

A recurrent theme in books and articles about corporate planning is that it is not a course of action one takes because business is bad. Indeed, business difficulties would probably be the worst reason for making a corporate plan. The best plans are made when one is, as far as possible, in control of events not when events are on top of one.

Nor should one make changes for the sake of change. Change can be stimulating but it can also be disruptive. It is better for organizations to evolve and develop to meet changing circumstances than to be 'reorganized'. Adaptation to environment is the biological term – evolution is better than revolution.

In this connection it is worth having the following quotation in mind: 'We trained hard . . . but it seemed that every time we were beginning to form in teams we would be reorganized. I was to learn later in life that we tend to meet any new situation by reorganizing; and a wonderful method it can be for creating the illusion of progress while producing confusion, inefficiency and demoralization' (Gaius Petronius AD 65).

From this it is apparent that people, at least, have not changed very much over the last 2,000 years. But even if it seems that mankind in general does not learn much from experience perhaps we can profit from the experience of other businesses.

Contractors have never been afraid to look over competitors' fences and learn from what they see so they should not be reluctant to do so when considering the approach to corporate planning which is best suited to their needs. What are the factors which they should consider?

Management has often been described as 'getting things done through people' so obviously the human factor should be the first to be considered. Corporate planning is, when all is said and done, only part of the management function.

A major problem of corporate planning generally has been that it has tended to be carried out from 'top down' so that there is always a risk that middle-management and those below in the organization structure will feel that the plan has been imposed upon them.

There is nothing more likely to produce a plan which will fail than failure on the part of the planner to involve, from the beginning, the men, whose job it will be to carry out the plan.

If the man who will be running the project has had a part in making the important decisions about the selection of assistants, the type of plant, and the methods to be used he has a vested interest in the success of the project and he will do his very best to see that the decisions to which he was party are proved to have been correct.

Town planners, too, have in the last few years come to appreciate the need to involve the public in making plans which concern their future environment. Town planning decisions are much more likely to be accepted by the public if there has been 'public participation' in the decision-making processes. Clearly the human factor is as important in corporate planning as it is in other types of planning. Corporate planning must be undertaken by all levels of management with participation by everyone from the beginning of the plan. Without their involvement they will not have the will to make the plan work and without the will any plan is doomed to failure.

Next to the human factor the time factor is probably the most important when preparing a corporate plan. Very few people have time to stand back and look at what they are doing; they are usually too busy doing it.

How many people in the normal contracting organization have time to devote to making a corporate plan in addition to doing their own job? And how many would admit to doing it if they had?

If time is not available from within the company one must look outside; to a management consultancy.

A management consultant should be able to provide the necessary staff to help a company undertake a corporate plan much more quickly than would be possible if the company relied entirely on its own staff.

The consultant will not, however, be cheap and he will need to call on the time of members of the company if he is to make proposals for a meaningful plan. We have already established the need for the plan to be the product of all concerned in its implementation. Somehow, the time will have to be found and it will have to be found at times which are acceptable to all. An hour or two after normal hours when people are tired and want to go home is simply not good enough.

Even if the in-company staff can, somehow or other, find the time, have they the knowledge necessary to make a corporate plan? Once again one finds oneself looking towards management consultants to provide the expertise.

An argument frequently advanced (albeit usually by management consultants) is that only someone not employed in the company can be objective. Employees, they say, have vested interests in certain aspects of the company's activities and will tend to suggest plans which will advance their own interests and careers.

This is certainly an aspect to watch but if the plans are developed by all sections of the company concurrently there should be sufficient natural checks and balances to prevent any particular sectional interests from getting out-of-hand.

Before we start work on our corporate plan we should decide whether or not we will employ a management consultant. If we attempt to summarize the advantages and disadvantages our scorecard will probably read something like the following.

In favour of employing consultants

(a) consultants can provide the necessary staff in order to undertake the plan as quickly as possible;
(b) they have knowledge and experience of corporate planning in which we are lacking;
(c) they should be more objective as they can stand outside the organization and 'see the wood for the trees';
(d) they should have no 'axe to grind' so there is less chance of one section of the firm being favoured rather than another.

In favour of doing it ourselves

(a) We are saved the cost of consultant's fees (although we will have some extra in-company costs);
(b) the plan will be our own and we will have a vested interest in its success;
(c) we know our own business inside out;
(d) there is less chance of there being change for its own sake (there are stories,

23

probably myths, of corporate planning consultants recommending open-plan offices, brightly painted, if the existing offices are 'individual' but recommending small separate offices if the existing office is open-plan).

Obviously it is not enough to count the advantages and disadvantages and come down on one side or the other. We must look most carefully at all the factors before making our decision. We might ask ourselves: 'having in mind our lack of corporate planning knowledge and all the other time commitments of our own personnel can we afford not to employ consultants?'

Nevertheless, an appreciable number of firms have successfully undertaken their own corporate planning. Assuming the managing director decides not to employ a consultant and that he asks the senior manager to act as the co-ordinator of the corporate plan giving him a reasonably free hand to work out the programme, decide on the methods to be adopted and report back. It is agreed that the company will seek outside advice as and when required.

If we take into account the factors which we have discussed above we will probably come to the conclusion that we need an approach which will enable us to involve as many of our staff as possible. The approach should require the minimum of specialized knowledge.

This will probably lead us to consider the closed-loop approach but at the same time ensuring that there is the sort of participative and objective discussion between the corporate top management and heads of departments, which is associated with the strategic review approach.

Both of these approaches were discussed above.

A particular advantage of the closed-loop approach is that it makes provision for 'the circular response'.

There is a tendency to think that there will only be a linear (direct) response to actions which we take. Certainly there may be a linear response but the response might well be, indeed almost certainly will be, circular.

Circular response can best be explained by example and Mary Parker Follett (1926) used a game of tennis to explain the concept. A serves. The way B returns the ball depends partly on the way it was served to him. A's next play will depend on his own original serve plus the return of B, and so on. Whenever one acts one starts something and behaviour precipitates behaviour in others.

One is in command of the situation until one acts – as soon as one has 'hit the tennis ball with the racquet' one has lost control of it until it is returned by the other player. It is not suggested that one should not act; rather that one should concentrate to the full before doing so.

Aristotle said: 'To succeed you must first ask the right questions.' A major part of the corporate planning tasks referred to above is asking the right questions.

The six tasks

Task 1

Record current strategy: The first of the tasks is to record the current strategy.

They are similar to the stages gone through in method study and work measurement: .select, record, examine, develop, install and maintain often remembered by the mnemonic SREDIM.

The first of the questions to be asked when recording the current strategy is:

Q.1(a) *What is our current strategy?* This is primarily a question which concerns top management but experience shows that interesting and useful answers are often obtained from the heads of departments within the company and as our corporate planning approach includes participation by all, the question can be asked at all management levels.

The answer should be kept short.

The current strategy statement might be the result of two or three 'stabs' by those concerned and if the participation exercise is to be meaningful the heads of departments will have involved their own staff from the beginning.

The next question is:

Q.1(b) *What kind of business does management want to operate?* The factors which would affect the answer to this question would be the desired return on investment, growth rate, share of market, stability, flexibility, character of the business and economic/market climate. There would be several parts to this answer.

The answers might be: A business which gives a return on investment at least equal to and preferably more than that obtainable elsewhere: We require a return on borrowed money in excess of the rate at which we borrow the money. We want to grow at, at least, the projected national growth rate and we would consider 5 per cent to 10 per cent per annum to be a realistic rate to aim for.

We want to increase our share of the market particularly having in mind the recent demise of one of our principal local competitors some of whose works in progress we have taken over at the request of the receiver. We want to keep the present character of the business because it is flexible in the local market and is reasonably responsive to changes in the market and economic climates but we accept the need for some diversification to provide a cushion against the government's use of the construction industry as a regulator of the national economy.

That is what we consider we want by way of business.

Now for the next question:

Q.1(c) *What kind of business does management feel it ought to operate?* The factors to consider here are management's concepts of social responsibility and obligations to shareholders, employees, community, competitors, customers, suppliers, government and the like.

In our business the shareholders are, let us suppose, the widow of the founder, one or two family friends and relations who put in money during the early days of the company's development, three directors (holding 40 per cent of the shares between them) and some long-standing employees who, together, hold 15 per cent of the shares. It is a private, limited liability company and shares must first be offered back to the company if a shareholder wishes to sell.

None of the shareholders are dependent on the shares for their livelihood and generally they have received better return on their money than they would by investing in a building society or similar institution. Our regular employees have in many instances trained with the company and spent all their working lives with it and the average length of employment is twelve years.

We provide a modest service to the community by meeting part of its needs and we make, as a company, modest contributions to the local community. We enjoy good relationships with our competitors and we do not knowingly do them harm. We pay our suppliers and sub-contractors regularly and usually within the agreed periods. Many of our customers have returned to us time and again.

As for the government, we work for the government occasionally, do not evade paying our taxes although we are careful to observe the learned judge's comment that we owe a duty to our company to avoid paying more taxes than we need.

At the risk of appearing complacent our answer to the last question is: We feel we ought to operate a business similar to our present business but rather larger and, perhaps, more diverse.

Task 2

Identify problems with the current strategy: This is the second task and the first question when tackling this task is:

Q.2(a) *Are trends discernible in the environment that may become threats and/or missed opportunities if the current strategy is continued?* During the thirty years which 'our' firm has been in existence it has learned the hard way that fluctuations in the economy are one of the greatest threats. Just when things look set for an expansion there is another recession and it is necessary to move into a lower gear.

Because of the nature of our business we have not been unduly threatened but we have been glad of our contacts and outlets outside of the building industry on one or two occasions.

There appears to be a trend towards design/build projects. Our organization is geared towards projects obtained by traditional means (mainly competitive tenders via architects' offices). We may be missing future opportunities. Perhaps links should be formed with building contractors with a view to future 'package' projects.

Q.2(b) *Is the company having difficulty implementing the current strategy?* Our strategy is for growth and this is hampered by lack of work in an industry frequently in recession. The demise of one of our competitors may help us in this respect but it is too early to say as yet.

The company is not diversifying as it should to compensate for the market restrictions in its main product line.

In short, the company is not having difficulty but it should be in a stronger position than it is.

Q.2(c) *Is the attempt to carry out the current strategy disclosing significant weaknesses and/or unutilized strengths in the company?*

Our weaknesses are probably:

1. lack of know-how in some business areas;
2. failure to recognize opportunities;
3. failure to innovate.

Our unutilized strengths lie in:

1. our inability to use our expertise because of lack of work due to political and economic constraints;
2. the borrowing facility which our assets make possible but which we have not exploited.

Q.2(d) *Are there other concerns with respect to the validity of the current strategy?* Not that we are aware, but perhaps we are not competent to answer this question?

Q.2(e) *Is the strategy no longer valid?* We believe that it is still valid.

While Task 1 might be largely (but by no means solely) for top management Task 2 should most certainly be undertaken by all levels of management working concurrently. A look at Fig. 2.4 indicates a parallel working approach which is an essential aspect of a strategic review, an approach which will continue throughout subsequent tasks. The contractor should not be too concerned if it exceeds the time scale shown in Fig. 2.4. Using its own resources the firm may take longer to carry out the progressive tasks.

Task 3

Discover the core of the strategy problem: To some extent the questions asked under this heading overlap those asked when undertaking the second task but we can develop our thinking in this third task.

Q.3(a) *Does the current strategy require greater competence and/or resources than the company possesses?* When answering Q.2(c) under the last heading we recognize that we lack know-how in some specialist fields although it does not appear that we lack other resources.

Q.3(b) *Does the current strategy fail to exploit adequately the company's distinctive advantage?* Again, a previous answer points to an answer to this question. We probably are failing in this respect. We like to believe that our failure is due to circumstances outside our control but . . .!!

Q.3(c) *Does our current strategy lack competitive advantage?* It probably does not lack competitive advantage but we may not be exploiting it to the full.

Q.3(d) *Will our current strategy fail to exploit opportunities and/or meet threats in the environment, now and in the future?* We are beginning to establish a line of thought that leads us to answer 'perhaps we are failing now and will fail in the future in this respect'.

27

Q.3(e) *Are the various elements of the strategy internally inconsistent?* We think not but there is perhaps a risk that we might lose our 'concentration' if we diversify in too many directions. There is plenty of evidence that an absence of good management cannot be compensated for by diversification but on the other hand Fig. 2.2 indicates that a business which provides alternative growth directions is less vulnerable than one which 'puts all its eggs in one basket'.

One must maintain a balance in these things and we think our business has proved itself sufficiently well-managed to master greater diversification and growth.

Q.3(f) *Are there other considerations with respect to the core of the strategy problem?* None that are immediately apparent to us.

Q.3(g) *What, then, is the real core of the strategy problem?* We have developed successfully over the years but we have reached a stage where we need to look at markets which are outside our experience. Our top management has the will to grow but lacks expertise in certain business areas.

Perhaps we have been too busy with day-to-day matters to make it possible for us to stand back and see the wood from the trees. The demise of our competitor recently has probably been the catalyst which has caused us to make a strategic plan.

Task 4
Formulate alternative new strategies:

Q.4(a) *What possible alternatives exist for solving the strategy problem?*

1. Growth; recently made easier by the demise of a major competitor;
2. development of a design/build division to undertake 'package deals';
3. speculative developments, possibly in conjunction with a finance house/insurance company or similar organization which would use the property to enlarge their portfolios;
4. find new markets to exploit;
5. develop our range of products and services.

Obviously the above list is by no means exhaustive but it at least suggests some possible alternatives.

Q.4(b). *To what extent do the company's competence and resources limit the number of alternatives that should be considered?* We have already established that we lack managerial competence in some areas in which we could diversify and all but 1 in the above list probably require the recruitment of managerial expertise if they are to be viable.

Finance does not appear to be a major problem when planning our strategy as we know we are low-geared but we could develop cash-flow problems if we do not exercise careful budgetary control.

Budgetary control is discussed in Chapter 7. We are well placed as far as other resources are concerned. Money has been described as being 'convertible' in that it can be used to buy other resources – even good people. Money is a versatile resource and as we appear to have sufficient assets to enable us to borrow money, the company should not be too restricted in its future activities by limitations of resources. It may be necessary to give some alternatives preference over others but with careful planning and budgeting we may be able to develop more rapidly than we would anticipate.

Q.4(c) *To what extent do management's preferences limit the alternative?* The question to consider here is 'are we in business for fun or profit?' Our answer may be crucial for our future development.

We have already satisfied ourselves that we need alternative growth directions in order to make us less vulnerable to market, political and economic changes so we do not consider ourselves to be limited by preferences.

Q.4(d) *To what extent does management's sense of social responsibility limit the alternatives?* Not at all. We consider that the scale of our activities will not be to the detriment of others. We may have passing feelings of guilt that our success contributed to some extent to the failure of our former competitors but we are sufficiently pragmatic to realize that it is a competitive world and our guilty feelings evaporate quickly. We have, after all, taken many of their employees onto our payroll.

Q.4(e) *What strategic alternatives are acceptable?* We have not as yet tested the viability of the possible alternatives and this factor may be a deciding factor but in principle all the alternatives are acceptable.

Task 5
Evaluate alternative new strategies: It is often very difficult to make an objective evaluation of alternative proposed strategies. How can/does one compare available alternatives? How can one quantify the possible affects when there are so many variables, imponderables and unknowns?

To some extent the examination and analysis of trends in market, economy, technology, politics and society which were discussed above can provide a basis for evaluation but experience has shown that hindsight is a much more exact science than prediction.

Management must use its best judgement and endeavour to minimize future risks but accept that business is risk-tasking and that if one dislikes heat one should stay out of the kitchen. Having said that one should ask the questions listed under this heading:

Q.5(a) *Which alternative best solves the strategy problem?* With the limited information available it is not possible to determine a best alternative but it may be possible to suggest a best match with the company's competence and resources which is Q.5(b). One would expect, for instance, that diversification into products or services closely associated with our activities would provide the best match although this need not necessarily be so. We must remember that one reason for diversification is to reduce our vulnerability.

Q.5(c) *Which alternative offers the greatest competitive advantage?* Here, again, we have insufficient information in our model to give a valid answer to the question but a diversification complementary to our main activity might give the company the greatest advantage.

Qs.5(d) and (e) follow on from Qs.4(c) and (d) and refer to management's preferences and sense of social responsibility. We will probably find differences of opinion between the members of the management group as to their individual preferences.

This is one occasion on which an outside consultant might be useful as without an independent adjudicator it may be the loudest or most plausible voice which might prevail.

The question of social responsibility would probably not overtax the team. Building does not usually present as many social/environmental/ecological problems as some other industries.

Q.5(e) *Which alternative minimizes the creation of new problems?* Any diversification will create new problems and the need to introduce new personnel (which seems likely from our earlier deliberations) will be bound to create problems.

Alternatives nearest in character to our existing business are most likely to minimize the creation of new problems. We should not, however, avoid creating new problems if the overall benefits to the company are worthwhile.

Task 6
Choose a new strategy.

Q.6(a) *What is the relative significance of each of the preceding considerations?* This is the penultimate question which leads to the crucial decision which arises from the final.

Q.6(b) *What should the new strategy be?* If we have followed the outline planning process set out in Fig. 2.4 we can see that this is one of the 'combined tasks'.

If our firm has adopted a management group approach to planning (and hopefully in due course running) the company the final decision will be the concern of all. We are not concerned with a major change of direction but even so growth and diversification are not to be undertaken lightly.

Let no one imagine that having determined our strategy, all will be accomplished without more ado. There are plans to make which will doubtless require reappraisal, indeed the closed-loop process makes provision for a circular approach.

We have, then, only just started on the endless circle or 'loop' of corporate progress.

Cynics have suggested that 'forecasting is very difficult, especially if it is about the future and that the moment you forecast you know you are going wrong; you just do not know when and in which direction' (PD–OR).

When forecasting and corporate planning it is necessary to accept that there are many factors over which one has no control. The most important aspects of corporate planning are that the planning process imposes a discipline on the planners and that they are thereby required to, at least, attempt to be masters of their future rather than merely react to events.

References

Buckner, H. (1974) *Director's Guide to Management Techniques*. Gower Press, London.

Drucker, P. (1974) *Management – Tasks, Responsibilities, Practices*. Heineman, London.

Follett, M. P. (1926) *The Psychological Foundations, Constructive Conflict in Scientific Foundations of Business Management*. Williams and Wilkins, Baltimore.

Gilmore, F. F. (1971) Formulating strategy in smaller companies, *Harvard Business Review*, May–June.

Humble, J. W. (1974) *Director's Guide to Management Techniques*. Gower Press, London.

Hussey, D. (1974) How to plan success, *Management Today*, London, November.

Lumsden, P. (1971) Business planning in construction, *Building*, London, February 12.

Maslow, A. H. (1970) *Motivation and Personality*. Harper and Row, London.

Townsend, R. (1970) *Up the Organization*. Michael Joseph, London.

Young, R. and Hussey, D. E. (1978) Corporate planning at Rolls Royce, *Management Today*, London, November.

Chapter 3

Knowing the market and obtaining work

Obtaining work is the first and in many respects most important task that the building contractor, specialist contractor, architect, quantity surveyor or other consultants have to perform. Without work, no matter how competent at his job the businessperson may be, he will not be in business. It is no accident that many firms have one or more persons employed largely on obtaining work. The contractor (and the word may be used in this instance to include all the 'consultants' referred to above because they are all under contract in one way or another), needs a client who has a need for his services. His order of priorities is:

1. a 'service' to offer;
2. a client for his service;
3. an acceptable price for his service;
4. construction of product (the service he provides).

When the standard form of contract is used, the contractor does not design his product. His order of priorities differs from other 'traders'.

Manufacturers of and traders in mass-produced goods such as motor vehicles, clothing, detergents and household goods, etc. generally manufacture a product to their own design on a speculative basis. They design and manufacture a product for which they believe there is, or will be, a demand (market). They must have confidence in their design and the price for which the product will be sold because considerable investment may be involved. Obviously careful market research is of primary importance.

For sub-contract work on building projects the specialist contractor does not require an extensive knowledge of marketing but he should be aware of approaches to marketing planning which are in general use. Whilst these are directed mainly to the mass-produced product industries they are also applicable to a greater or lesser extent to the building industry. Market planning starts with determining the information which the specialist contractor needs to know about the market, economic, political and legislative factors and about his competitors' activities.

Information requirement for market planning

In order to prepare a marketing plan the contractor needs to know[1]:

A. About the market:

1. Total size of the market.
2. Rate of change of the size of the market.
3. Factors which affect changes.
4. Number of competitors.
5. Competitors' share of the market.
6. Competitors' change in market share.
7. Competitors' sales organization structure.
8. Ruling current market prices.
9. Forecasts of market conditions.
10. Seasonal/cyclical market fluctuations.
11. Market potential (new and existing).
12. User characteristics/attitudes/opinions.
13. Potential customers – kind, number and location.
14. Uses of service or product.
15. Customers service/product selection criteria.
16. Sources of customer dissatisfaction.
17. Competitive position of firm's service.
18. Distribution methods.

B. About economic factors:

1. Industrial cycles.
2. Profits, profit cycles and trends.
3. Share prices.
4. Order/sales cycle.
5. Investment cycle.
6. Level of unemployment.
7. Trends of economic activity.

C. About political factors and legislation:

1. Regional development plans.
2. Ecological pressures.

D. About competitors' activities:

1. Competitors' strengths and weaknesses.
2. Attitudes to firm price and/or fixed price policies.
3. Servicing policies.

4. Type of distribution/selling cover.
5. Promotion activity.
6. Policy to debtors/creditors leasing.

Many of the above items do not at first sight appear relevant to the specialist contractor but they should always be used as a check-list when reviewing the market. Occasionally, a closer look at check-list reveals that an item has greater relevance than at first appeared to be the case.

The contractor should, for instance, have in mind the rate of change of the size of the market (item A2) when considering whether or not to expand his production capacity. The change of the size of the market would, in turn, be influenced by cyclical market fluctuations (item A10) and trends of economic activity (item B7).

Reference is made to these cycles and fluctuations in Chapter 1. The specialist contractor may consider it prudent to avoid heavy investment in plant for production when all indications are that demand is at its peak and likely to decline.

Market outlets and media

Of more immediate and practical importance to the contractor would be item A13 – potential customers – kind, number and location.

His principal customers would initially, probably, be architects and consulting engineers but if present trends indicate a growth in, say, speculative housing developments his range of potential customers could be extended to include house-builders.

Marketing may be defined as 'market-orientated planning' so it is not concerned with implementing the plan. Contractors are often reluctant to 'sell' themselves because they prefer to rely on the recommendations of established clients. But despite their reluctance, contractors should accept the increasing tendency towards commercialism which extends into the professions and regard marketing and selling as serious management functions.

There has been an increase in the use of marketing agencies by contractors and certainly these agencies are able to provide expertise in a field which is alien to contractors. Before launching into professional marketing and selling, however, specialist contractors may consider using the obvious means of promotion such as the local and national press, telephone directory 'Yellow Pages', the technical and professional press, personal canvassing of potential clients, stands and trade fairs, presentations at branch meetings of professional institutions, sponsoring conferences, local and national television, local radio, etc. Many media are available for the specialist contractor who is seeking to expand his sphere of activities.

Following up tenders submitted

One marketing medium which surprisingly few contractors exploit fully is the tender for work which has been submitted by the contractor/sub-contractor. A prospective 'client' (who for a sub-contractor may be a main contractor) is, to at least some extent, under an obligation to a firm which has invested the considerable amount of time and money which is involved in submitting a tender (see Ch. 4). The tenderer has, therefore, a significant psychological advantage which should be exploited.

The tenderer should always follow up the tender he has submitted to discover how it compared with those of his competitors and in order to develop a business rapport with the prospective client.

Sub-contractors who discover that the project, for part of which they submitted a tender to main contractor A, is likely to proceed with main contractor B, to whom the sub-contractor did *not* submit a tender, may find it advantageous to inform main contractor 'B' that he has submitted a tender.

Following up in this way may be providing the contractor with a new contact or a closer relationship with an established client, and gives the contractor at least some reward for the high cost of tendering even if the tender itself is not accepted.

The market which is available for specialist contractors is a crucial factor which he must consider when deciding his aims, objectives and corporate plan and reference is made to marketing in Chapter 2, but before he attempts to plan his future he should appreciate the fundamental nature of trading activities.

Fundamental trading activities and the need for profit

For most practical purposes trade and business comprise one or a combination of two or more of the following activities:

1. The purchase of materials and/or goods at a given price and their sale at a (usually) higher price. (This activity includes wholesale and retail trading.)
2. The use of labour to provide a service. (Unskilled and skilled labour, the crafts and the professions all come in this category.)
3. The purchase or 'winning' of material and the addition of labour so that the material's form is changed. (The manufacturing industries, including the building industry, may be included under this heading.)

All the above activities add value to whatever existed previously – hence the concept of value added tax (VAT). The added value provides a profit (the value of the newly created 'whole' exceeds the sum of its parts), which is necessary for the business to grow, for it to innovate and in order to provide a cushion against the risk which is inherent in any business activity. Furthermore, taxation on profits from business activity pays for the social services which any civilized community expects its government to provide.

A law of the market-place in which business activity is carried out is summed up in the economic theory of supply and demand. Simply expressed, this runs that for a commodity (or a service) to have a value there must be a demand for it. As the supply of the commodity rises to meet demand its value decreases until, when supply exceeds demand, it has a little or no value.

For an object/service to be of value it must:

1. Be in demand.
2. Meet the purchaser's need (conform with his specification).
3. Be within the price which he is willing and able to pay.
4. Be available when he needs it.

The contractor/sub-contractor ignores these rules at his peril. Without a demand he will not be in business and unless he observes rules 2, 3, and 4, he will not *stay* in business.

Note

1. These lists have been adapted from Shaw, W. C. and Day, G. J. (1978) *The Businessman's Complete Checklist*. Business Books, London.

Chapter 4

Estimating and tendering

' "Estimating" is the technical process of predicting costs of construction.

"Tendering" is a separate and subsequent commercial function based upon the estimate' (CIOB 1979).

Tendering – final step in the marketing process

Tendering is usually the final step in the process of obtaining work. The aim is to submit a price which will cover the cost of carrying out the work and provide a surplus – a profit. The principal constraint on the upper limit at which a contractor tenders is 'what the market will stand'. Tendering is ultimately therefore a marketing decision.

An elementary example to illustrate the marketing principle would be a subcontractor being invited to submit a tender to install replacement windows at No. 64 Acacia Avenue which resulted from the owner of No. 64 having inspected windows recently installed in No. 62.

Assuming the neighbouring houses were similar and the specification for the windows to be installed was similar to that for No. 62 the contractor's tender would be based on:

(a) manufacturing cost of windows plus;
(b) installation cost plus;
(c) management cost plus;
(d) profit.

In the example, (a) and (b) would be known from the contractor's costings (cost feedback) from No. 62's installation. His factory would provide records, probably in the form of a computer printout, for (a). His installers' timesheets, probably similar to the daily allocation sheet shown in Fig. 10.1 (p. 160), would enable him to calculate the cost of (b). If his installers had been working on a 'piece-work' or labour-only sub-contract basis there is little doubt but that he would have been informed by them if their price for No. 62 had put them at a loss so one way or the other the estimator would know the cost of both the manufacture and the installation of the windows for No. 62. So much for the technical estimate of cost comprised in (a) and (b). The contractor's calculations of his overhead cost, the cost of running his establishment – factory, office rent, rates, staff, etc. which make up (c) would probably be expressed as a percentage to be added to (a) and (b). The actual percentage to be added would be calculated on the basis of an anticipated annual turnover using one of the methods described later in this chapter. 'Turnover' is the value of work carried out during the year. It is frequently called 'sales' by accountants.

Element (d) of the contractor's tender (his profit) is that which requires commercial judgement. Profit is necessary as a cushion against the risk which any business activity entails and for innovation and growth but the actual margin to be included in the tender will be influenced by numerous factors.

In this example, the prospective customer's 'expectation' will probably be taken into account when deciding on the profit margin to be added to the estimated cost of the work. The owner of No. 64 will no doubt have spoken with No. 62 and so he will be aware of what No. 62 paid for his windows. Number 64 would not be likely to approach the contractor for an estimate if he was not prepared to spend a similar sum. It follows, therefore, that even if the contractor discovered that the cost of (a) and (b) for No. 62 were less than he estimated when preparing his original tender he might think twice before giving the owner of No. 64 a tender which was *less* than that given to No. 62. He may reasonably argue that the owner of No. 62 might be dissatisfied if he later discovered that No. 64 had paid less than he for essentially the same work.

On this occasion the contractor might consider submitting a tender to No. 64 with a similar price to that given to No. 62 (having made allowance for inflation, etc. which had occurred since the tender for No. 62 was submitted) and accept that he should, assuming no unforeseen snags, obtain an enhanced profit on No. 64: the market conditions are, in this instance, in the contractor's favour.

If, on the other hand, the contractor discovered from his records that the cost of (a) and (b) for No. 62 had exceeded his estimate but competition from other contractors and shortage of work for his factory made it important for him to obtain more work in order to avoid making men redundant, he might well *reduce* his profit margin when tendering for No. 64. In each case the contractor is influenced when making his final decision by the market conditions – what will the market stand?

The above example is elementary but it demonstrates the principles of tendering: a tender is a means of obtaining work which comprises: (a) a *technical estimate* of the cost of the work to be carried out and; (b) a commercial decision regarding the profit margin which should be added to the estimated cost in order to ascertain a tender sum which will be acceptable to the client and also reimburse the contractor for his enterprise.

With this general picture of estimating and tendering in mind one may look in more detail at the various stages in the process.

Stages in estimating and tendering process

Estimating and tendering for work comprises nine separate stages:

1. an invitation from a client to a (specialist) contractor to tender for work;
2. a decision by the contractor to submit a tender;

3. quantification of the work to be carried out;
4. decisions regarding the methods to be used when carrying out the work;
5. estimating the cost of the labour, plant, and materials to be used;
6. ascertaining the duration of the work;
7. estimating the cost of the firm's 'overheads' attributable to the work;
8. making a decision regarding the desired profit margin having in mind commercial and market factors;
9. submission of tender.

The invitation to tender

An invitation from a prospective client as to a contractor's willingness to submit a tender for work is frequently the result of the contractor's efforts to exploit the market (see Chs. 1 and 3) or it may arise from the recommendation of a former satisfied client. The contractor has cause to be particularly pleased when invitations are from the latter source.

The invitation may take the form of a letter or telephone call asking if the contractor would be interested in tendering for a particular job in the near future or the first indication may be the arrival of a parcel containing drawings, specifications or bills of quantities. Specialist contractors are occasionally invited to advise on the type of construction or installation which might be suitable for a project as a prelude to a more detailed invitation.

The first decision which faces the contractor is whether or not he wishes to tender. His decision will be influenced by his corporate plan (see Ch. 2) and his available resources. Manpower and cash availability are usually the principal factors which affect his decision.

If the contractor is disposed to decline the invitation but he believes such an action may prejudice his chances of being invited to tender on future occasions he may decide to accept the invitation but take steps to ensure that his tender will not be accepted.

This may be done by submitting an inflated tender. The risk that the contractor takes in this event is that his tender will appear so unrealistically high that he will lose his credibility with the client.

Cover prices
To avoid this risk contractor 'A' may ask a competitor to suggest a sum which he, contractor 'A', may use as his tender sum. The competitor will obviously suggest a sum which is in excess of his own tender sum but contractor 'A' will be confident that the tender sum he will be submitting will appear to be competitive.

Giving and taking 'cover prices', as the practice described above is called, is not in itself illegal but it is discouraged in the building industry because it may lead to illegal collusion.

Opinions vary regarding the reactions of clients when their invitation to tender is declined. Most clients maintain that they would prefer contractors openly to decline an invitation and that such an action does not prejudice the contractor's future. Nevertheless, many contractors remain unconvinced in this respect and cover arrangements are not unusual.

Decision to tender

The decision to tender is a decision to incur cost because estimating requires knowledge, experience, skill and time. With experience a contractor can make a shrewd 'guesstimate' of the cost of a job without numerous calculations and if the profit 'margin' he can allow himself in his estimate is large enough he can make a few errors on the 'swings and roundabouts' principle without coming to any harm.

If market conditions are favourable and the contractor is in a position to include a profit of, say, 20 per cent then ± 5 per cent is neither here nor there. If, however, his profit margin is only 10 per cent an error of 5 per cent is much more significant and could make a difference to his expansion plans for next year. The contractor should, therefore, take care with his estimate. It is the foundation on which his success rests. Obviously, no contractor can afford to spend too much time on tenders because the odds against him getting it are, perhaps, 5:1, but he should remember that if his tender is accepted it makes sense if his estimate can be converted into his production plan.

A methodical approach to estimating

The estimate may be of importance to the contractor not just during but after the works have been completed and the final account is being prepared. An estimator can only take into account the information which is available to the contractor at the time of tender. If the works are varied during the contract period from those for which the contractor tendered he is entitled to have the contract sum varied. Evidence of the information which was available for him at the time of tender may be invaluable when agreeing the cost of variations or claims.

The contractor should, therefore, take care when preparing his estimate to record all the documents used in its preparation, so that his estimate will be intelligible, perhaps, several years later. Care should also be taken to ensure that the estimate details are not mislaid during the course of the project. Much may hang on the contractor being able to prove that this or the other piece of information was or was not known at the time of tender.

If, then, the contractor decides to submit a tender he should plan his programme for the preparation of his estimate and the submission of his tender with care. A methodical approach is an essential ingredient of estimating. This chapter is concerned primarily with principles and method because the technical aspects of diverse specialist works are beyond the scope of a book of such a general nature but the principles of tendering are common to all types of work which the specialist sub-contractor may be called upon to undertake.

Quantifying the work

An estimator must know the quality and quantity of the work for which he is preparing an estimate. For the traditional building project the quality will normally be specified by the designer so the estimator has no decisions to make in that respect.

For some specialist work the contractor may be provided with bills of quantities prepared by the quantity surveyor but frequently the estimator must take-off (measure) the quantities himself. In either event it is true to say that for all but the simplest projects it is necessary to take-off the work to be carried out. Indeed, subconsciously, a contractor 'quantifies' the work even when giving an estimate for a simple job. It is the *units* in which contractor or estimator thinks which vary from job to job. In the replacement window example given earlier the contractor estimated by 'the house' – both houses being similar. For a more complicated project he would need to think in more detailed terms.

Much building work is measured in accordance with the current edition of Standard Method of Measurement of Building Works (SMM) which is authorized by agreement between the RICS and NFBTE after consultation with various trade associations.

The SMM contains twenty-three sections, the majority of which are 'work' sections with headings such as demolition and alterations, excavation and earthworks, piling, asphalt work, plumbing and engineering installations, electrical installations, glazing, etc.

The SMM has suffered considerable criticism over the years but it has the advantage that it provides a generally understood basis for tendering even if for analytical estimating purposes it is often necessary to group some of the items in the work sections into discrete operations.

If, then, a bill of quantities has been provided for the specialist contractor's use when submitting his tender he will be saved an appreciable amount of taking-off. He may consider it necessary for him to re-group some items in the bill of quantities or break-down individual items into, say, separate floors, carcassing and 'second fixings', etc., depending on the way he envisages planning his operations, but he will, at least, have an indication of the total extent of the work which has to be carried out.

If bills of quantities are not provided the contractor will find it necessary to take-off his own quantities.

The scope of the work is a major determinant of the method of taking-off to use. For major works dimension paper is probably to be preferred. Identical items of work which occur on various sheets of dimension paper may be abstracted on to the reverse side of the method statements for the appropriate operations. Method statements are discussed below.

For some contractors a standard ruling, often described as 'estimating paper', may be appropriate. This enables the contractor to take-off dimensions in the left-hand columns and price the items in the right-hand columns using one column for labour and the other for materials.

41

Analysis		Rate	Labour	Plant	Material	Total
TILE ROOFING						
B of Q 60 (e)						
In tile roofing fixed						
to a 45° pitch ——— per m²						
The number of 265mm × 165mm						
tiles per m²						
Gauge = length of tile lap						
2						

		Rate	Labour	Plant	Material	Total
length of battens per m₂						
$= \dfrac{1m}{\text{tile gauge}}$						
$= \dfrac{1m}{95mm} = 10\cdot53\,m$						
Say 11 m/m²						

		Rate	Labour	Plant	Material	Total
Tile outputs						
tiling — 100 tiles/hr						
battening — 50 m/hr						
tiling = $\dfrac{64\,tiles}{100\,tiles/hr}$ = 0·64 hrs/m²						
labourer = 0·32 hrs/m²						
battening $\dfrac{10\cdot53\,m/m_2}{50\,m/hr}$ = 0·21 hrs/m²						
labourer = 0·11 hrs/m²						

Fig. 4.1 Estimating analysis

Another format is shown in Fig. 4.1. This ruling is useful for analysing operational costs from items contained in bills of quantities and so building-up unit rates for sub-contractors' tenders to clients, architects or main contractors from the labour, plant and materials rates which should form the basis for all 'combined' rates.

The method used to ascertain the quantity of work and the actual form of presentation is unimportant provided the estimator is able to use the data to calculate the cost.

The technical estimate

Technical estimating is concerned with predicting the cost of production. Successful construction and/or installation consists of bringing together the ingredients of production at the right time and in the right place and at the right price in order to meet the client's need. The ingredients being:
(a) manpower directly employed by the contractor;
(b) plant;
(c) materials and manufactured components;
(d) sub-contractors;
(e) management.

The last is the catalyst which aids the 'chemical change' in the other bodies.
Estimating requires:
(a) knowledge of production methods (which in practice is usually derived from observation and experience);
(b) an ability to analyse from observations and synthesize new methods to be used for the proposed works;
(c) an ability to ascertain the cost of the new methods.

For most specialist contractors' purposes the estimator is concerned with:
(a) how many man-hours will be required?;
(b) which plant is to be used and for how long?;
(c) which material will be required and how much (or many) of each?;
(d) what managerial, supervisory and administrative back-up is needed?

The method to be used

The total estimate should be built up by means of *method statements* of the *operations* necessary to complete the project. The sub-contractor's *project* is unlikely to consist of one operation. Most usually it consists of a number of operations, some of which, at least, are isolated from the others by work which has to be carried out by other contractors.

Indeed, an *operation* in a building context is often described as a piece of work which can be started and finished by a man or group of men without interruption by the work of others. An operation is, in effect, a 'work package'.

A method statement is simply a statement, usually quantified, of the way in which the operation will be carried out; how the job will be done, and the type and quantity of the ingredients needed for its execution. It is the link between the estimate, the production plan and the work itself should the tender be successful.

When the estimator has decided on the operations which he will use for the preparation of his estimate he is in a position to analyse the method which is most appropriate for each operation and so estimate his resource requirements: the manpower, plant, materials, etc. which will be necessary to carry out the operation.

Method statement format

Figure 4.2 provides an example of a pro forma which may be used when preparing a method statement. The format is such that it may be used, first, for the estimate, and later, provided the tender is accepted, for assisting management with its production planning and control.

The top section of the pro forma makes provision for the project title and the specialist contractor's own reference number or code to be inserted. The reference EOM/80/11, is derived from the project title Excell Offices, Middleton. The letters are simply a short form of the title which can be used for all references to the contract within the contractor's organization for costing, marking packages, records, ordering, etc.

Because the initial letters readily relate to the title the reference quickly becomes accepted by personnel in the contractor's organization generally and provides an easy means of identification. The number '80' is the *year* when the job first entered the contractor's office and the last numbers show that it was the eleventh job that year.

As any enquiry is received it is automatically logged. Because the contractor will only undertake the successful tenders there will be a considerable number of 'missing' numbers but in practice this does not matter.

Reference was made earlier in the chapter to the need for the estimator to work in a methodical manner and to ensure that all relevant information is preserved. In this connection, he may find the quantity surveyor's method useful for marking papers. Because quantity surveyors may at any one time have a considerable number of dimension sheets, abstracts, etc. relating to several jobs floating around the office, they use a child's *printing set* to make a stamp for marking all papers.

Below the title on the method statement pro forma is space for a description of the operation to which the method statement refers. This need not be detailed as far as the *method* is concerned – detail can be worked out and given under heading 3f of Fig. 4.2.

It will be noted that the *operation number* has not been inserted. This is because at estimate stage, the stage we are concerned with at present, the number is not known. When the contractor arranges his operations in order to make his pre-tender plan he can pencil in an operation number but usually he will not commit

METHOD STATEMENT for contract ...Wellington Road Housing.. ref: WRH/80/8.........

Operation No.   for   External walls and structural partitions – DPM to First Floor including seating for precast planks

as drawing and specification refs: 290/A/F/1–4,C/F/1 and 3

1. BQ Spec. Ref.	2. Materials a. Description	b. Quant.	c. Unit	d. Rate (£)	e. Cost (£)	f. Or-dered	g. On site	3. Labour and Plant a. Unit rate	b. Unit hrs.	c. Total hrs	d. Man-days	e. Cost (£)	f. Gang/plant description
5/10/J-H	Artificial stone sills												Call in labour rates: Craftsman £2.60/hr Labourer £2.30/hr
	800 Long	48	nr	6.20	297.60			1.40	0.54	25.92		67.20	
	1600 "	32	"	11.30	361.60			2.20				70.40	
	ends	160	"	0.25	40.00			0.10				16.00	
5/12 H-N	Lintels												
	1,100 Long	48	"	4.70	225.80			2.00				96.00	
	1,900 "	32	"	7.90	632.00			2.70				86.40	
5/15/A	Expanded polystyrene	80	m	0.30	24.00			0.15				12.00	
5/13/C	Rawlbolts	180	nr	0.35	63.00			0.30				54.00	
5/17/B	Facing brickwork 1½ piers	20	m²	18.20	364.00			10.40				208.00	
C	Corbelling	8	m²	22.50	180.00			12.10				96.80	
5/18 A-D	Labours		Item					200.00				200.00	
5/19 B-J	Facings ditto 1B	72	m²	11.30	813.60			9.00				648.00	
	Labours		Item					62.00				62.00	
5/20A-P	Facings ditto ½B	1,870	m²	7.70	14399.00			5.50				10285.00	
	Labours		Item					370.00				370.00	
	Lignacite 220	1,230	m²	6.20	7626.00			3.80				4674.00	
	Labours		Item					290.00				290.00	
	Dacatie 50	360	m²	1.40	504.00			0.60				144.00	
				—	£ 2,9622.35							£ 18962.40	Labour
		90	m					0.65				£ 58.50	Profiles 100x50 sawn

Fig. 4.2 Method statement

45

himself to a number until he prepares his production plan.

Care should be taken to enter the revision letters of the drawings, in addition to the numbers, because the architect or consulting engineer is probably developing the design at the same time as the estimator is preparing his estimate and the drawings which the estimator is using are probably already obsolete. It is essential that the estimator marks all the drawings 'Estimate Drawings' near the date stamp mark (all documents entering the contractor's office should be date-stamped as a matter of course) and that the drawings are carefully preserved with all the papers relating to the project estimate.

Heading 1 may be used for references to the bills of quantities or for reference to item numbers or pages in the specification or in the specialist contractor's own take-off.

Section 2a can be used to summarize the items contained in the take-off. Although the costs of all the items have been collected in section 2e in the example in Fig. 4.2 it is preferable to include the rates of all items in section 2d.

Sections 2f and 2g are irrelevant at tender stage but they are useful for processing the ordering of the materials and controlling their delivery to site. The dates in these columns can be invaluable for *instant reference* when chasing deliveries or establishing that an order *has* been placed and that a variation at this stage really *will* involve cancellation, compensation etc.

If the cost of the operation has been built up using unit rates for the items which comprise the operation, the labour content of the rate is inserted in section 3a. It may be that specialist contractors do not use unit rates very much when compiling their estimates but the column is there for their use if they need it.

Calculation of the *total hours* and man-days are largely a matter of arithmetic using the *all-in hourly rates* which were used when compiling the estimate. Such rates are contained in the 'Schedule of Labour Cost' published from time to time in the technical press. The schedule takes into account the basic rate paid to an operative, the JIB benefit stamps, CITB levy, and provides for special rates on large sites, London Weighting and country/lodging allowances.

The number of man-days is calculated:

$$\frac{\text{cost of operation (col. 3e)}}{\text{all-in-hourly rate} \times \text{hrs/working days}}$$

If a craftsman/technician and assistant will be working as a 'pair' one should read pair-days for man-days, above. The duration of the operation can, of course, be shortened or lengthened by increasing or reducing the number of men or pairs as the case may be. Reference is made to this later in this chapter when discussing pre-tender planning.

Section 3 makes provision for both labour and plant. The plant which would be included here would be that which would be used just for this operation. If, for instance, the sub-contractor had to supply a mobile tower to obtain access to his work the cost of such plant would be included in section 3.

The method statement is, then, a 'tool' which is fashioned at estimate stage and used for planning, coordinating and controlling the works if the tender is accepted.

Figure 4.2 is the method statement for one of a number of operations which go to make up the works at Excell Offices and all the operations must be integrated into a pre-tender plan.

Materials, labour and plant costs

For most practical purposes materials and components are the easiest of the ingredients' costs to estimate. The materials should be scheduled so that the full specification, complete with BSS references and the quantities of the materials, may be circulated to suppliers. Care must be taken to ensure that the suppliers of brought-in components are provided with sufficient information, preferably in the form of drawings, sketches, etc. to enable the supplier to estimate the cost of manufacturing the components.

Specialist contractors are well advised to have prepared a standard form of enquiry for suppliers' materials which should be formulated to take into account the contractor's requirements. Such pro formas prevent items being overlooked. Most should, at least, include the following items:

(a) the specification of the material;
(b) the quantity of the material;
(c) the delivery programme – period during which supplies would be needed and daily or weekly requirements;
(d) notice required by supplier for 'calling up' materials if deliveries are to be phased;
(e) address of site;
(f) requirements regarding packaging, palleting, provision of lifting eyes, etc.;
(g) means of access to site, delivery time restrictions; as appropriate;
(h) period/s for which estimate is to be 'firm price' and/or open for acceptance;
(i) person within contractor's organization whom supplier should contact regarding queries.

Contractors will normally obtain quotations from several suppliers. Care should be taken when comparing estimates for competitiveness to ensure the quantity of each item is taken into account as well as individual rates. A number of low rates for items for which the contractor will require small quantities of the materials may be more than offset by one low rate for which a large quantity of the materials will be required.

When pricing materials for estimating purposes, the estimator should make allowance for materials wasted, damaged or 'lost' during the course of the works.

Labour costs

Labour costs are the most difficult ingredient of production to estimate. Even when a person is performing a routine, repetitive task on a production line in a factory it is not always possible accurately to estimate the time he will take to carry out his task. Under variable, difficult-to-control building-site conditions the time taken by a person to carry out a non-repetitive or semi-repetitive task may vary considerably from project to project or even in different parts of a single building site. Estimating the cost of labour for an operation to be carried out under unknown conditions cannot be an exact science.

Work measurement and method study
Production line estimating is traditionally based on work measurement and method study. The work or method selected for study is recorded and examined and a new method is designed, developed and installed. Over a period of years the movements involved in various activities have been measured and analysed so that 'standard' times are on record from which it is possible to synthesize a standard time for any new activity. Synthetic times may be used for estimating on building projects but for the reasons stated above they are of limited value.

Obtaining estimating constants
Another method of work measurement which is generally more appropriate for building projects takes a broader study of the labour involved in an operation by recording the *whole* time taken on the operation and then calculating the cost of each unit within the operation. An elementary example of this would be to take the recorded time for an operative installing, say, ventilation trunking for 'one floor' in a building. Assuming that there were 150 m run of trunking and that it took 50 man-hours to install it, the unit time would be

$$\frac{150 \text{ m}}{50 \text{ man-hours}} = 3 \text{ m/man-hour}$$

Statistically it would be necessary to make more than one measurement before reliable data would be available for the estimator but over a period of time a contractor may obtain a library of 'constants' for use when estimating for future work.

The time taken to carry out individual operations may be recorded on an allocation sheet of the type shown in Fig. 10.1 (p. 160) and the method of recording is described in Chapter 10.

Calculating all-in labour rates
Having ascertained the man-hours for a unit or operation it is necessary to calculate a labour rate which takes into account wages and all associated costs using the all-in rates referred to above.

Plant costs

The role of plant in specialist contractors' work varies greatly from contractor to contractor. Sub-contractors are not normally expected to provide plant for transporting their materials to the point of installation in the building but they should take care when estimating to ascertain their obligations. Chapter 5 discusses the action which the sub-contractor should take in this respect.

Sub-contractors for whom plant is an essential ingredient of production may obtain 'output' constants from records in the manner described under the previous heading.

Hired plant costs
These may be estimated by taking the cost of hiring the plant for the duration of the operation and adding the cost of fuel and operator if these are not included in the plant hire cost.

Sub-contractors' plant costs   These are more difficult to estimate because the sub-contractor will not usually know the actual useful life of the plant, the possible cost of maintenance, the proportion of the life of the plant when it will actually be 'producing' (not standing idle in the sub-contractor's yard) etc. Records are useful as a means of making realistic estimates. Daily allocation sheets as shown in Fig. 10.1 (p. 160) may be used to guide the estimator as to the productive use to which the plant is being put. Contractors are often surprised by the low proportion of the life of a piece of equipment during which it is actually 'earning'.

Having analysed records the calculations for obtaining hourly or daily running costs is a matter of arithmetic.

The manner in which plant cost is included in the sub-contractor's estimate will depend, largely, on the part it plays. If, for example, the plant is primarily supportive and will be on site for the whole duration of the sub-contractor's work, it may be impractical to allocate the plant cost to individual operations. A scaffold tower or crane supplied by the sub-contractor for his own use would come into that category. The cost of plant used in the above manner would be estimated and included as a lump sum or, if it were considered appropriate, the lump sum cost would be calculated and added to other site overhead costs and included in the tender as a percentage addition to the unit rates. Overhead costs are discussed below.

Integrating labour, plant and materials costs in the estimate

Having measured the quantity of materials and components required for an operation, having estimated the time required by the appropriate operatives to carry out the work involved in the operation by means of synthesis or constants and having assessed the plant requirements the estimator is able to calculate the estimated cost of the operations.

Allowances for site conditions

The estimator's technical skill, experience and judgement will be exercised to 'measure and assess'. It must be remembered that 'constants' are averages and the estimator should make allowance for any conditions which may affect the manner in which his work would have to be carried out in the event of his tender being accepted.

If the project has not commenced the estimator should take into account the competence of the design team in so far as he has experience of it or is able to obtain information about it.

The tender procedures which have to be followed to comply with the basic method described in SF 80 (see Ch. 5) should provide the sub-contractor with indications of some of the conditions he should expect *vis-à-vis* the contractor.

The estimator should study the drawings to ascertain the extent of the work and, equally important, to satisfy himself that the project has been designed in such a way that his work may be carried out without its regular progress being unduly disturbed or varied. The competence record of the architect and consultants should be kept in mind.

If the project is in hand the estimator is better placed to ascertain factors which could affect the way in which the sub-contractor may find it necessary to carry out the work in the event of his entering into a contract.

Site visit

A visit to the site is important. The estimator will be able to see for himself site conditions, transport and lifting facilities and the efficiency of the contractor's site management if the works are in progress.

The estimator should prepare a check-list or pro forma of items which he should note when making a visit to ensure that the distractions which inevitably accompany a site visit do not lead him to overlook important items. The items on a check-list would vary from one specialist sub-contractor to another. A piling or earthworks contractor, for example, would regard information concerning the nature of the subsoil and the level of the water table to be of paramount importance whereas such information may not interest a building services contractor. The estimator should regard the check-list shown in Table 4.1 as indicative and extract from it the items which apply to his type of work. A permanent record of conditions and information available at the time of his visit is important to the estimator for the preparation of the estimate and for post-contract purposes.

From what has been said above it will be apparent that estimating is more than the use of labour and plant constants and materials rates. Useful as they are, labour and plant constants should be regarded as a starting point from which adjustments should be made to take into account site conditions.

The cost of labour which is calculated using the pro forma format shown in Fig. 4.1 is, then, a sum calculated for the operation as a whole which may be divided by the number of units contained in the operation if the sub-contractor is required to submit his tender in the form of unit rates. Whilst the unit rate will generally combine labour, plant and material costs it is important for production planning and control purposes that the sub-contractor has, in the first place, built up his unit rate in the manner shown in Fig. 4.1.

Table 4.1 Indicative check-list for use when making initial visit to site at time of estimate

1. Project Title .... Job Ref. ....
   Address of Site ....
   .... Visit by .... Date of visit ....

2. *Communications*
   2.1 Nearest town
   2.2 Nearest bus station
   2.3 Nearest railway station
   2.4 Nearest public transport stops
   2.5 Available tips for refuse/spoil disposal

3. *Local authorities*
   3.1 District council
   3.2 County council
   3.3 Job centre
   3.4 Water/sewerage undertaking

4. *Site access*
   4.1 Adjoining road. Condition/width
   4.2 Local obstructions (low/weak bridges, etc.)
   4.3 Temporary roads required into site
   4.4 Temporary roads required on site
   4.5 Possible access difficulties

5. *Site description*
   (Provide plan/sketch with details if possible)
   5.1 Existing roads to use/demolish/reinstate
   5.2 Existing buildings to demolish/preserve
   5.3 Existing foundations/obstructions
   5.4 Existing trees, hedges to remove
   5.5 Existing trees, hedges to preserve
   5.6 Existing surface water
   5.7 Adjoining buildings
   5.7.1 To be supported
   5.7.2 Obstructing site activities
   5.8 Ground conditions
   5.8.1 Topsoil
   5.8.2 Sub-strata
   5.8.3 Bore-holes (existing/required)
   5.8.4 Water-table level
   5.8.5 Springs/tides
   5.8.6 Pumping required
   5.8.7 Earth support required

6. *Services availability*
   6.1 Drainage: foul/surface water
   6.2 Water
   6.3 Gas
   6.4 Electricity: voltage/3-phase supply
   6.5 Telephone

Table 4.1 cont'd

7. *Security on site*
    7.1   Fencing: existing/required
    7.2   Watching
    7.3   Sheds/compounds
    7.4   Regulations

8. *Labour*
    8.1   Local rates of pay
    8.2   Local customs (bonus/plus rates, etc.)
    8.3   Availability locally
    8.4   Need to import

9. *Plant*
    9.1   Local suppliers/hirers
    9.2   Servicing facilities
    9.3   Haulage

10. *Materials*
    Local suppliers

11. *Sub-contractors*
    Local sub-contractors

12. *Main contractor's administration*
    12.1   Site transport
    12.2   Cranes/hoists
    12.3   Scaffolding/towers
    12.4   Site management

13. *Other information*
    13.1   Drawings for reference
    13.2   Specification/schedules
    13.3   Other relevant sources

Assessing contract duration

The estimator needs to know the anticipated duration of his work in order that he will be able to calculate the cost of supervision and similar items, the costs of which are related to the time his work will be in progress.

To estimate the work period the estimator must prepare a programme which he is able to assemble from his method statements. The planning process is discussed in Chapter 7 but for pre-tender purposes it is sufficient to say that estimating and planning should go hand-in-hand.

Planning

A *basic* plan is essential in the preparation of an estimate. Conversely a production plan which ignores the estimate is likely to produce a financial loss. The production plan should be determined by:

(a) the resources which are available (which should have been allowed for in the estimate);

(b) the technology of the project;

(c) the most economic method of deploying the resources.

The most simple way of explaining what this means is to take as an example an operation which it has been calculated will take 100 man-hours to complete. The manpower resource can be deployed by using:

```
  1 man  for 100  hours
  2 men  for  50  hours
  4 men  for  25  hours
  5 men  for  20  hours
  8 men  for  12½ hours
 10 men  for  10  hours
 20 men  for   5  hours
 25 men  for   4  hours
 50 men  for   2 hours
100 men  for   1  hour
```

In theory the result should be the same but in practice there is usually an optimum manpower strength for an operation. Too few men on the job and the supervisory costs are increased and are spread over a long period and the job loses its impetus. Too many men and productivity can suffer because the men are 'falling over each other'. Too few, or too many, and the sub-contractor's work can become out-of-phase with the main contractor's work to the detriment of both. If, or rather when, the contractor's tender has been accepted the estimate can be converted into the production plan. The production plan takes the resources which the contractor has allowed in his estimate, as set out in his method statements, and determines the optimum time for the project and the optimum manpower strength at any stage in the works.

From the planning standpoint the sub-contractor may have the overall period in which he is to carry out his work determined for him by the main contractor or he may be appointed before the main contractor in which event he will have freedom to decide for himself, within reason, how he plans his work.

To illustrate the alternative circumstances take as an example a project for the construction of Excell Offices with an electrical contractor undertaking the electrical installation as a nominated sub-contractor.

If the main contractor has provided a programme at the time the electrical contractor submits his tender, the electrical contractor's contract duration will be largely determined for him and he can price his management and supervision item at an appropriate rate. The main contractor for Excell Offices *has* provided a programme (see Fig. 4.3), which allows the electrical contractor 17 weeks during which to carry out the carcassing work, 4 weeks for fixtures and equipment and 4 weeks during which to test and commission the installation. Three periods of five weeks separate these work periods.

The electrical sub-contractor has allowed in his estimate for a supervisor

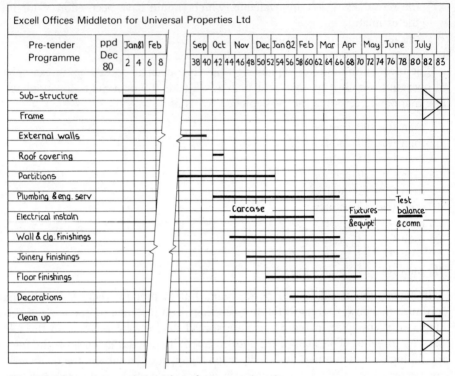

Fig. 4.3 Main contractor's bar chart for pre-tender plan

spending $2\frac{1}{2}$ days and the contract engineer 2 days each week on the Excell Offices during the 25 working weeks. The supervisor has been priced at £60/day and the engineer at £70/day.

It must be remembered that although the main contractor has specified in his programme the *periods* during which the electrical sub-contractor is expected to carry out his work, the sub-contractor's men may not necessarily be on the site during the whole of the period. If they are not, the sub-contractor will have over-priced the site management and supervision content of his tender but it is better to err on the safe side.

If the main contractor has not provided a programme the sub-contractor will have to prepare his own and this leads us to look at the elements of planning.

The principal aims of planning are:

(a)   to determine a completion date;
(b)   to determine which operations are the most important (critical) and those which can be delayed without altering the overall completion period (non-critical operations);
(c)   to determine the sequence of operations;
(d)   to assist the contractor to identify his information requirements (drawings and schedules from architect and engineer, main contractor's dates, suppliers' delivery dates, etc.);

(e) to provide a basis for controlling work in progress and;

(f) to provide a basis for cash flow and financial control.

It is vital that the plan is flexible and capable of being revised quickly and easily. Building projects, however well organized, are vulnerable to changes and sub-contractors are less than realistic if they do not accept this and plan their work accordingly.

Bar-charts

An extract from a bar-chart is shown in Fig. 4.3. The walls of most building contractors' site offices are decorated with such charts. This form of presentation has been used for production planning purposes for many years but was not adopted by the building industry in general until after the Second World War. Gantt, a management pioneer, gave his name to these charts and for projects with comparatively few operations they are often perfectly adequate. Bar (Gantt) charts also have the advantage that they are readily understood by the people who need to use them. A disadvantage of these charts is that they do not necess- arily show the dependency of one operation upon another.

A main contractor will normally prepare a pre-tender plan along the lines of that shown in Fig. 4.3 which covers the principal stages in the project such as when it will be 'out of the ground', when it will be 'watertight' and when it will be completed. It also provides information about the entry and exit dates of the principal sub-contractors. It is not much more than indicative but the main contractor will usually have calculated his plant durations (how long the crane will be needed on site, etc.) from such a chart.

The programme is, of course, very simple and needs little explanation. The principal object is to place the operations on a time-scale. At pre-tender stage the operations are much larger units than those used for short-term programmes. Operations can often be broken down into smaller units and that is what the electrical sub-contractor has done where, as may be seen in the tender summary illustrated in Fig. 4.4, he has identified fourteen operations which he will slot into the three bars on the main contractor's programme.

Similarly, the time-scale will vary with the type of programme being prepared. For pre-tender plans the units can be *weeks* as in Fig. 4.3.

As will be seen in Chapter 7, shorter units may be used for more detailed programmes.

Sub-contractor's expectation from main contractor

Whilst ascertaining the duration of the sub-contract works the sub-contractor should make a note of the attendance and the facilities which he expects the main contractor to provide.

The interface between the main contractor and the *nominated* sub-contractor in respect of attendance and facilities is described in Chapter 5. The points made in that chapter should provide a guide to the sub-contractor estimating for both

Excell Offices, Middleton -- Tender for electrical installation

Operation	Materials	Labour	Plant	Net totals	Total incl 10% o/heads	Total incl 5% profit	Tender sum
1 Mains switchboard and meters	604.00	102.40	–	706.40	777.04	815.89	
2 Rising mains and cables	2,310.00	205.60	105.00	2,630.60	2,893.66	3,038.34	
3 Distribution boards and connections	1,270.00	980.00	120.00	2,370.00	2,607.00	2,737.35	
4 Lift mains and controllers	365.00	120.20	22.00	507.20	557.92	585.82	
5 Plant room mains and connections	1,290.00	1,780.20	225.00	3,295.20	3,624.72	3,805.96	
6 Fix only convectors	215.00	490.00	50.00	755.00	830.50	872.03	
7 Sanitary incinerators and connections	270.00	215.60	90.00	575.60	633.16	664.82	
8 Fix only incinerators	36.00	52.00	–	88.00	96.80	101.64	
9 Fire Alarm Wiring	1,200.00	1,130.00	140.00	2,470.00	2,717.00	2,852.85	
10 Fix only equipment	360.00	430.20	35.00	815.20	896.72	941.56	
11 Air-conditioning plant and connections	2,360.00	2,170.00	130.00	4,660.00	5,126.00	5,382.30	
12 Lighting/wiring/outlets	6,720.00	4,890.40	270.00	11,880.40	13,068.44	13,721.86	
13 Under-floor ducts	5,980.00	4,620.80	310.00	10,910.80	12,001.88	12,601.97	
14 Clock installation	1,630.00	1,680.00	110.00	3,420.00	3,762.00	3,950.10	
15 Site supervision and management	–	–	–	7,250.00	7,250.00	7,612.50	
Prime cost sums (as list attached)	24,610.00	18,867.40	1,607.00	52,334.40	56,842.84	59,684.99	59,684.99
	–	–	–	26,400.00	26,400.00	27,720.00	27,720.00
Cash discounts (add 1/39th)	–	–	–	78,734.40	83,242.84	87,404.99	87,404.99
							2,241.15
Tender sum							89,646.14

Note: This column is used to make adjustments to the estimate immediately prior to submission of the tender at the discretion of the directors.

Fig. 4.4 **Tender summary**

nominated and domestic contract works and the schedules contained in tender NSC/1 (see Fig. 5.3, 5.4 and 5.5, pp. 70–3) should be used as check-lists.

'Firm price' estimates

The majority of tenders are prepared using rates for labour, plant and materials which prevail at the time of tender. The estimator realizes that the work, if the tender is accepted, will be carried out *in the future* and that provision must be made for increases in cost which may occur between the time when he prepares his estimate and when the materials will be purchased, the plant hired, or the actual work carried out.

Labour increases will usually be known in advance (at least a few months ahead) because the trade unions concerned negotiate increases with the appropriate employers' federation. The estimator may use his pre-tender plan as a means of assessing the volume of work which should be carried out before and after the date when proposed wage increases will take effect.

Materials and plant increases are often difficult to predict. If the tender is for works of relatively short duration the sub-contractor may persuade his suppliers and plant hire firms to give him 'firm' prices but they will seldom do this for work more than a few months ahead. If firm prices are not forthcoming the estimator must make his own prediction of inflationary trends, perhaps, using one of the indexes which are published. Here, again, his pre-tender plan should assist him to assess the volume of work which will be completed each month so that he can make an appropriate allowance. The sum assessed is normally included in the tender as a lump sum.

Fluctuation estimates

For major projects, clients usually appreciate that it is impractical for contractors to submit firm-price tenders for work to be carried out over a period of years ahead. It has become common practice for 'fluctuation' or 'variation of price' tenders to be submitted for works which it is anticipated will take more than one year to complete or which will not commence in the immediate future and which will, therefore, be difficult to price due to uncertainty regarding commencement and completion dates. The Standard Form of Building Contract contains provision for fluctuations (SF 80, Cls 39 and 40 and NSC/4, Cls 34–37).

Overhead costs

Overhead costs are those costs which are not attributed to the cost of production but which are, nevertheless, part of the cost of running the business. They are often separated into 'direct' and 'indirect' overhead costs.

Direct overhead costs

These are costs which relate to a particular project. They include items such as site management and supervision, site accommodation and temporary services and plant of the supportive type referred to above. Direct overhead costs may be calculated on a weekly or monthly basis which may be converted to the total cost for the contract by relating the weekly/monthly cost to the estimated duration of the contract. The total cost of the direct overheads may be included in the tender as a lump sum or as a percentage addition to the unit rates in the manner suggested when referring to supportive plant costs above.

Indirect overhead costs

Such costs are those which relate to the contractor's 'establishment'. These costs include his office rent, rates and the pay of his permanent staff such as secretarial support, payroll and accountancy staff, telephone operator, etc. – the staff who will have to be paid regardless of the firm's level of activity. Overhead costs also include the cost of financing the firm. If the firm is running on an overdraft or loan, the interest payable on it should be paid, in part, by each project. Alternatively, if the firm is financed by the proprietor's or partners' own funds they are entitled to reimbursement at, at least, the rate of interest they could expect by investing their money elsewhere. Financing costs include the cost of financing the project itself.

Indirect overhead costs may be estimated for the forthcoming year/s on the basis of previous years' costs.

Each project should contribute a share towards the indirect overhead costs. Without the support of the firm's establishment there would be no contract! The costs are usually allocated to individual contracts pro rata the value of the firm's turnover. If, for example:

Estimated turnover for the year is	£500,000
Estimated prime cost of a contract is	£ 25,000
Share of overhead cost for the contract would be	

$$\frac{25,000}{500,000} \times 100 \ = 5\%$$

If the estimated overhead cost for the year is	£ 45,000
the share for the contract would be 5% × £45,000	£ 2,250

Alternatively, and perhaps preferably, the overhead cost may be expressed as a percentage of the turnover and used to calculate the sum to be allocated to the contract:

$$\frac{£45,000 \text{ estimated overhead cost}}{£500,000 \text{ estimated turnover}} \times 100 \ = 9\% \times £25,000$$

$$= £2,250$$

A contract's share of indirect overhead costs should be built into the tender so that it may readily be identified should the contractor's tender be successful.

The reason for requiring clear identification is that the estimate should be the *budget sum* for an ongoing project and clear identification of the cost components makes it simpler for the actual production costs to be monitored against the budgets (see Ch. 7).

Although the methods used to calculate overhead costs which are described above are quite widely ₘused by specialist contractors they are really only directly applicable if the firm's turnover and overhead costs are unlikely to change greatly from one year to the next. Clearly, if turnover is increased without a corresponding increase in overhead costs the ratio of overhead costs : turnover will be reduced and the increased scale of activities will provide economies. Conversely, a reduced turnover without a corresponding reduction in overhead costs will make the firm less competitive in simple estimating terms.

There is, then, some flexibility of approach to overhead costs and a contractor keen to obtain work may be prepared to tender for work with a smaller addition to his production (prime) cost for overhead costs than the calculations outlined obove would indicate as being appropriate. He cannot, however, do so lightly or for a prolonged period without inviting insolvency.

Profit margin – the commercial decision

In the replacement glazing example given above reference was made to the variability of contractors' profit margins. It was suggested that the profit would be 'what the market will stand'.

Theoretically, numerous factors affect the margin such as demand for the product, competition, etc., some of which are discussed in Chapter 3.

In practice only a few firms attempt a scientific analysis of the profit margin which they hope should secure the project – just.

Analysis of competitors' profit margins

The cartoon in Fig. 4.5 illustrates the 'intuitive' approach to profit margin calculations. Experience has shown that time spent by the estimator in recording his firm's (company A's) tender data and comparing it with the tender sums of company A's competitors enables him to make better informed predictions of profit margins which may secure future contracts. Most contractors are aware that all too often the successful competitor submits a tender which appears to be uneconomically low; occasionally less than the figure which the estimator believes to be his firm's prime cost. No amount of comparative analysis of tenders will assist the competitor to compete with 'idiot' tenders; the contractor must hope that their folly will remove them from his list of competitors. But analysis of the 'sensible' competition frequently indicates that company A could increase its profit margin by several per cent and continue to submit competitive tenders. Analysis of competitors' profit margins makes it possible for the contractor to obtain some indication of 'what the market will stand'.

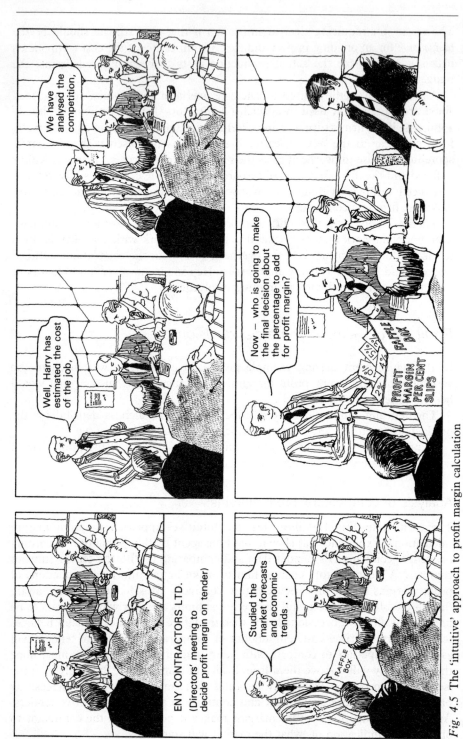

Fig. 4.5 The 'intuitive' approach to profit margin calculation

One approach to analysing tendering efficiency is put forward by William Park (Park 1978).

The basis of his approach is that in a competitive tendering situation the highest price that a contractor can set for his work is the lowest price that his cheapest competitor is willing to take to do the job.

If this is so, tendering efficiency may be defined as the ratio of the amount of profit actually made to the amount . that *could* have been made had all the competitors' tenders been known in advance. Alternatively, it represents the amount which could have been made by taking all the jobs at the lowest competitor's price, *provided the jobs would have been wanted at those prices*. The concluding proviso is significant and should not be ignored.

To adopt Park's approach the contractor should keep a record of his tenders with the essential data tabulated under the headings shown in Table 4.2.

Table 4.2 Record of tenders

Project no.	No. of competitors	Lowest competitor's tender (£)	Estimated cost (£)	Tender sum (£)	Maximum profit potential (£)	Actual profit (£)
1	2	95,400	87,600	96,300	7,800	0
2	5	41,400	36,300	40,600	5,100	4,300
3	4	416,500	388,300	428,900	23,200	0
and so to						
9	5	478,900	457,500	498,500	21,400	0
			1,884,000		107,100	26,700

Assuming nine recent projects on four of which the contractor submitted the lowest tender, estimated direct costs total is £302,800 and a profit of £26,700 assuming the cost estimates were correct. His additions for profit on the nine projects varied from 5 to 14 per cent. The lowest competitor's tender for each project determines the maximum profit that can be made on that job so that the 'maximum profit potential' represents the difference between the lowest competitor's tender and the estimated project cost.

Therefore, the maximum profit potential defines the most profit that could possibly have been made even if all the competitors' tenders had been known in advance for these nine projects, the maximum profit potential was £107,100. The £26,700 profit actually realized by the contractor on the four projects he carried out represents 24.9 per cent of his maximum profit potential on the nine projects:

$$\frac{£26,700}{£107,000} \times 100 = 24.9\%$$

A tendering efficiency within the range of 20 to 30 per cent is typical of many contractors who operate intuitively.

A table may be constructed showing the effect of the contractor applying different additions for profit is shown in Table 4.3.

Table 4.3

Percentage addition applied to project	No. of projects won	Estimated cost of projects won (£)	Gross profit on projects won (£)	Tendering efficiency (%)
0	9	1,884,000	0	0
1	9	1,884,000	18,800	17.5
2	8	1,250,900	25,000	23.3
3	8	1,250,900	37,500	35.0
4	8	1,250,900	50,100	46.7
and so to				
12	3	170,900	20,600	19.2
15	1	22,400	3,400	3.2
20	0	0	0	0

Between the extremes at 0 and 20 per cent addition for profit, different addition will produce, for example, £37,500 at 3 per cent and £20,600 at 12 per cent. A low percentage addition will produce more projects without, necessarily, a corresponding increase in profit, as the table shows. In the example given above a 7 per cent addition would win sixth-ninths projects and produce a profit of £54,500 but an additional 1 per cent addition would reduce the profit to £31,300. At 7 per cent addition the contractor would have a tendering efficiency of 50.9 per cent. There is clearly scope for experimentation and a larger sample of projects would produce more statistically significant results.

Break-even calculations

This method of calculating profit margins is more appropriate for production-line manufacturing than for 'unique' product manufacturing of the type undertaken by many specialist contractors. Nonetheless, some specialist contractors exist, primarily, because they act as an adjunct to a production line so the method is not inappropriate.

For purposes of break-even method, costs are considered under the headings of *overhead costs* and *unit costs*.

Overhead costs include the items discussed earlier in the chapter. If one uses the replacement window contractor as an example, overhead costs would include the cost of running the factory and the office costs.

Unit costs would include labour, materials, fuel and costs for each unit produced.

The overhead costs plus the unit costs comprise the expenditure cost which may be measured against the *income* or *revenue* from the sale of the product/s to ascertain the point at which the expenditure and revenue break-even.

The mathematical solution is not difficult but a graphical solution makes it easier to demonstrate trends on which to base management decisions.

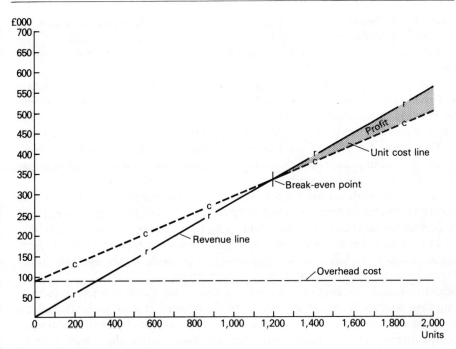

Fig. 4.6 Break-even analysis

The graph shown in Fig. 4.6 assumes that the overhead cost for the period is £90,000 which is shown as a horizontal line. The *period* used for the graph may be the day, week, month or year. In Fig. 4.6 the *cost for the year* has been used.

The unit cost line has its point of origin where the vertical axis meets the overhead cost line. In the graph it is derived from an assumed unit cost of £208.33 per window. The revenue line is plotted using the sale price of each unit of sale. The sale price is determined by market conditions. In this example the sale price is assumed to be £283.00. In the graph, revenue will break even with expenditure when 1,201 units have been sold and from that point forward the contractor will make a profit. An advantage of the graphical presentation is that management, knowing the number of units which must be produced before a profit will be made, is able to plan its sales and calculate profit margins on the basis of costs. For products which have a relatively low unit cost and a relatively high overhead and development cost – books, for example – the unit sale price may be greatly reduced when sufficient copies have been sold – hence the 'remainders' on offer in railway station shops and street markets.

Submission of tender

When the estimator has estimated the cost of the labour, plant and materials to be used and ascertained the duration of the work the management team, proprietor or board of directors (as the case may be) will normally become involved in the decision-making process prior to submission of the tender.

To facilitate decision-making a tender summary similar to that shown in Fig. 4.4 should be prepared.

Items 1–4 are calculated from method statements. Item 15, site supervision and management costs, is based on the duration of the works. The management team would become involved when the 'net totals' column had been calculated. The team would decide on the cost additions to be made for overheads and profit.

The 'Prime Cost Sums' total takes into account all the items of equipment which the architect has specified should be provided by nominated suppliers and for which he has stated a prime cost (PC) sum in the specification.

Provision is made for the 2.5 per cent cash discount which the main contractor is entitled to deduct from the sums certified for payment to the nominated sub-contractor by the architect in accordance with the conditions of contract (SF 80).

The column on the right of the tender summary headed 'Tender Sum' provides space for last-minute adjustments to be made to the estimate totals. Estimates are almost invariably prepared under pressure. There is seldom sufficient time and suppliers and sub-sub-contractors are frequently late in submitting their tenders to the sub-contractor. The sub-contractor's estimator often has to prepare his estimate on the basis of the information available only to find more competitive materials estimates arriving only hours before the tender is to be submitted to the client. Such contingencies may be taken into account in the tender sum column.

The example given above has assumed that the sub-contractor is required to submit a lump-sum tender and that fully-priced bills of quantities are not required by the client. When priced bills of quantities are required the estimator must calculate the unit rates which he is required to insert against each item in the bills of quantities.

Opinions vary regarding the methods which should be used when pricing bills of quantities. Some contractors maintain that each unit rate should include part of the overhead and profit and others that the last-mentioned items should be included in the tender as separate items. This issue is of importance when variations occur and for that reason it is discussed in Chapter 12 which is concerned with variations.

References

CIOB (1979) *Code of estimating practice*, CIOB, Ascot.
Park, W. R. (1978) *Construction Bidding for Profit*. Wiley, Chichester.

Chapter 5

Sub-contract tender and nomination procedures

For some 900 years English records of building works show that for most of that time all but the smallest projects were built by gangs of specialist craftsmen: masons, carpenters, plumbers, etc. who were employed direct by the person commissioning the building works (Salzman 1967). Occasionally these gangs were paid by the hour or day but frequently they undertook to carry out the masonry, carpentry or plumbing works for an agreed sum – they were 'specialist contractors'.

Early in the nineteenth century there was a great increase in the number of 'general contractors' undertaking building works. These contractors were usually building craftsmen with an entrepreneurial flair who contracted with the client to undertake *the whole of the building works*, not just the work entailed in their own craft.

The general contractor employed individual craftsmen of all crafts and he also employed gangs of craftsmen as 'sub-contractors'. General contractors were in existence long before the nineteenth century but a number of factors combined to cause the early nineteenth century escalation in their use.

At that time architecture became established as a profession. Architects usually encouraged their clients to employ general contractors but the architects frequently wished part of the work to be carried out by the craftsmen with a reputation for good workmanship. Forms of building contract were devised which made it possible for architects to nominate specialist contractors as sub-contractors.

There were, then, two ways in which sub-contractors were appointed: by the contractor and by the architect. This was referred to as the 'present-day arrangement', in Chapter 1 (see Fig. 1.1, p. 3). The 1980 edition of the Standard Form of Building Contract introduced a variation from this arrangement which is referred to below.

Alternative approaches to sub-letting

Nominated sub-contractors frequently become involved with an architect for a new project before the main contractor. There are three alternative ways in which

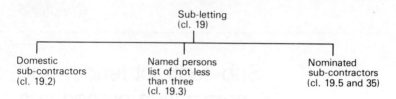

Fig. 5.1 Alternative approaches to sub-letting

a main contractor may sub-let work and the whole question of sub-letting is set out in clause 19 of SF 80. These alternatives are shown in Fig. 5.1. There is no reason why all three types of sub-letting shown in that figure should not be used on any one contract and the specialist sub-contractor may find himself employed under any of the methods.

Domestic sub-contractors (with architect's consent)

The main contractor 'shall not without the written consent of the architect sub-let any portion of the works' (cl. 19.2) but subject to that proviso, which appeared in earlier editions of the JCT form, there is nothing to stop the main contractor employing any sub-contractor as a 'domestic' sub-contractor to carry out specialist works which are measured in the bills of quantities. It will be interesting to see if there is an increase in the number of bills of quantities which actually have the specialist works measured. The new forms do not, then, introduce significant changes in this respect.

Named persons

Under clause 19.2 the architect takes no part in the selection of the sub-contractor – the choice is the main contractor's provided the architect consents to the selected sub-contractor. Clause 19.3, however, introduces an innovation which states that where the contract bills provide, certain work measured or otherwise described in those bills and priced by the main contractor *must be carried out by persons named in a list* in, or annexed to, the contract bills and selected therefrom by, and at the sole discretion of, the main contractor. 'The persons' in the list are named by the architect and the list must comprise not less than three persons (cl. 19.3.2.1). It is a typical British compromise. The architect keeps overall control of the firms (persons) who will act as sub-contractors but the main contractor is also able to exercise some control over the final selection because it is his 'say' which decides whether sub-contractor A, B or C is employed. At least, that is the position in theory but in practice one would imagine that the vast majority of main contractors will select the *lowest* tender as the sub-contractor's tender will be incorporated in the main contractor's tender and the main contractor could lose the contract on price if he selected too many sub-contractors whose tenders were not the lowest.

Nominated sub-contractors

Reference to the nomination of sub-contractors appears in clause 19.5.1 of SF 80 and this simply states that the provisions of the contract relating to nominated sub-contractors are set out in Part 2 of the conditions.

Part 2 of the conditions is concerned with both nominated sub-contractors and nominated suppliers. Clauses 35 and 36, which make up Part 2, to all intents and purposes replace clauses 27 and 28 in earlier editions of the Standard Form of Building Contract between the employer (client) and main contractor. At a recent conference introducing the new documents one of the members of the working party which compiled them said that the working party envisaged an increase in the volume of nominated sub-contract work in the 1980s. It is suggested that the reasons for nominating sub-contractors which were outlined by the Banwell Committee in 1963 still hold good (Banwell 1964):

(a) the architect requires *special techniques* to be used; techniques which are the province of a specialist (sub-)contractor;

(b) it is important to place an order for specialist work at an *early date*, probably before the main contractor has been selected;

(c) a *particular quality* of work is required which is the province of a specialist.

Basic method

There is a *basic* method and an *alternative* method of nomination which can be used by the architect. It is generally agreed, especially by sub-contractors, that the *basic* method should be used for the vast majority of contracts and especially where major specialist works are concerned or where the sub-contractor makes a significant contribution to the design of the works. The 'operations' to be carried out are shown in Fig. 5.2 in network form. The numbers which appear below the line are the relevant sub-clauses in SF 80. Follow through the operations to ascertain the sequence of events before considering the detailed implications.

Operation 1–2, 'architect selects sub-contractor'. Selection can be made before or after the main contractor has been appointed. Having selected the sub-contractor the next two operations run, largely, concurrently.

Operation 2–3 requires agreement from the sub-contractor to eventual nomination and involves completion of tender NSC/1 which consists of twelve pages. The aim of NSC/1 is to provide a means of negotiation between architect, sub-contractor and main contractor, so that when a contract is eventually entered into all parties have reached agreement on the terms.

Operation 2–4 requires the employer and sub-contractor to make an agreement so that the sub-contractor will be paid for design work, etc. should he subsequently *not* enter into a contract with the main contractor, (see op. 8–9), and so that the employer may use the sub-contractor's design if the sub-contractor is not employed to carry out the sub-contract works.

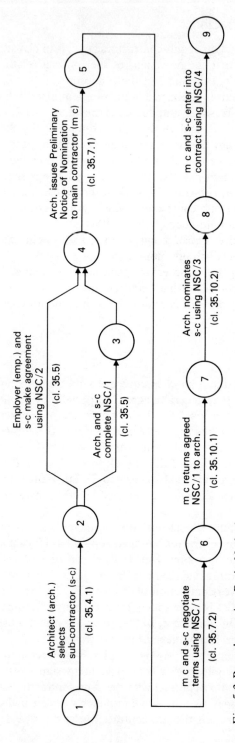

Fig. 5.2 Procedure using Basic Method

Completion of operations 2–3 and 2–4 make it possible for the architect to issue a Preliminary Notice of Nomination to the main contractor (op. 4–5) which is really an expression of the architect's intention to nominate the sub-contractor provided the sub-contractor and main contractor reach agreement as to details of attendance, contract periods, etc. which they attempt to do in operation 5–6.

Once terms have been agreed the main contractor returns NSC/1 (which will probably look rather like a dog's dinner by this time for reasons which will be gone into later) to the architect (op. 6–7) which enables the architect to nominate the sub-contractor using NSC/3 (op. 7–8). The, now, nominated sub-contractor and main contractor enter into a contract using NSC/4 (op. 8–9).

NSC/1, 2 and 3
Completing tender NSC/1

Having considered the sequence of events one may look in detail at the procedures using the 'operations' contained in the network, (Fig. 5.2), as the basis.

Selecting the sub-contractor (op. 1–2)
The JCT *Guide* points out that whilst the architect *may* use form NSC/1 for obtaining competitive tenders, or for negotiating with one sub-contractor, its use is not compulsory. Any documents may be used for the *selection* of the sub-contractor. Once, however, selection has been made the architect and sub-contractor must complete NSC/1 which, together with sub-contract NSC/4, will eventually comprise the principal sub-contract document.

Completing NSC/1 (op. 2–3)
A complete specimen of p. 1 of NSC/1 is reproduced as Fig. 5.3. From the printed wording on p. 1 it can be seen that either a lump-sum quotation ('VAT – exclusive Sub-Contract Sum') or a unit rate quotation ('VAT-exclusive Tender Sum') with detailed quantities and prices may form the basis of the sub-contract. The parties should take care to ensure that any documents which have been used in calculating the sub-contract or the tender sum are clearly numbered and identified.

Page 1 also has space for daywork percentages to be entered. These percentages are to be added to the 'prime costs' which are defined by the negotiating bodies; the RICS, NFBTE, ECA, HVCA, etc.

When the architect and sub-contractor are satisfied that the 'sum' and percentages which have been entered on p. 1 represent their intentions they sign at the foot of the page. It will be remembered that agreement NSC/2 is signed by the employer and sub-contractor at the same time as the architect and sub-contractor sign NSC/1 (op. 2–4 in Fig. 5.2). Further reference is made to NSC/2 later.

The stipulations on page 2 of NSC/1 are important. The first explains the relationship of NSC/1 and NSC/2. It reads: 'Only when this tender is signed on Page 1 on behalf of the Employer as "approved" and the Employer has signed

JCT

JCT Standard Form of Nominated Sub-Contract Tender and Agreement

See "Notes on the Completion of Tender NSC/1" on page 2.

Main Contract Works:[a] Community Centre for
Worsten District Council
Worsten Causeway, Worsten,
Location: BEDS.

Job reference: CCW/29/82

Sub-Contract Works: Electrical Installation

To: The Employer and Main Contractor [a]

We _J. Shine Electrical Services Ltd._

of _10 High Street,_

Nearbigh, Beds Tel. No: _0123-45678_

offer, *upon and subject to the stipulations overleaf,* to carry out and complete, as a Nominated Sub-Contractor and as part of the Main Contract Works referred to above, the Sub-Contract Works identified above in accordance with *the drawings/specifications/bills of quantities/schedule of rates for the Sub-Contract Works which are annexed hereto, numbered

Drawings Nos. CCW/1D, 2B, 3B, 4C, 7A Specification pp. 90-96 incl.

and signed by ourselves and by the Architect/Supervising Officer; and the Particular Conditions set out in Schedule 2 when agreed with the Main Contractor; and JCT Sub-Contract NSC/4 which incorporates the particulars of the Main Contract set out in Schedule 1.

*for the VAT-exclusive Sub-Contract Sum of £ _26,720.00_

Twenty six thousand seven hundred and twenty pounds (words)

~~or for the VAT-exclusive Tender Sum of~~ £ (words)

The daywork percentages (Sub-Contract NSC/4 clause 16·3·4 or clause 17·4·3) are:

Definition*[c]	Labour %	Materials %	Plant %
RICS/NFBTE			
RICS/ECA	100	15	15
RICS/ECA (Scotland)			
RICS/HVCA			

The Sub-Contract Sum/Tender Sum and percentages take into account the 2½% cash discount allowable to the Main Contractor under Sub-Contract NSC/4.

Signed by or on behalf of the Sub-Contractor _J Shine_ director for J Shine Electrical Services Ltd. Date 14th August 1982

Approved by the Architect/Supervising Officer on behalf of the Employer _AD Signer_ RIBA Date 22/VIII/82

ACCEPTED by or on behalf of the Main Contractor subject to a nomination instruction on Nomination NSC/3 under clause 35·10 of the Main Contract Conditions _A. Driver_ for P. ROPER BUILDERS LTD. Date 17 Sept. 82

© 1980 RIBA Publications Ltd *Delete as applicable Page 1

Fig. 5.3 Form of agreement, NSC/1

70

or sealed (as applicable) the Agreement NSC/2 do we agree to be bound by that Agreement as signed by or sealed by or on behalf of ourselves.'

The sub-contractor and architect may well sign on page 1 before the identity of the main contractor is known. Stipulation 2 reserves for them the right within fourteen days of written notification by the employer of such identity to withdraw the tender and the agreement NSC/2 notwithstanding any approval of the tender by signature on page 1 on behalf of the employer. In stipulation 3 they reserve the right to withdraw the tender if they are unable to agree with the main contractor on the terms of schedule 2 of the tender.

The tender is withdrawn if the nomination instruction (nomination NSC/3) is not issued by the architect within an 'acceptance period' which is inserted in stipulation 4 *by the sub-contractor*.

The final stipulation on page 2 is that withdrawal under stipulations 2, 3 and 4 shall be at no charge to the employer except for any amounts that may be due under agreement NSC/2.

Pages 3–12 of NSC/1 comprise schedules 1 and 2.

The provisions of schedule 1 are designed to set out the main contract terms which are then part of the nominated sub-contract. Schedule 1 should reflect accurately the terms of the main contract executed between the employer and main contractor. This is particularly important where the architect and the proposed sub-contractor have completed, as far as possible, the tender section of NSC/1 before the main contract has been settled and executed.

The sub-contractor should pay close attention to the insertions which are made in schedule 1 because it is only the main contract terms which are notified in this schedule which are binding on him. There is reference to this in NSC/4, the form used by the main contractor and sub-contractor in operation 8–9 when they execute their contract. NSC/4 is the 1980 edition equivalent of the former Green Form conditions. There are two appendices to schedule 1 which are concerned with fluctuations and one or the other will be deleted depending on the method of adjusting variations of price which is to be used.

Schedule 2 of NSC/1 is titled 'Particular Conditions' and it is this schedule which provides the main contractor and the sub-contractor with a means to negotiate the conditions which will eventually be included in the contract between the two contractors.

In the words of the 'note' to schedule 2 the main contractor 'has to complete this schedule in agreement with the proposed sub-contractor. The completed schedule should then take account not only of the preliminary indications of the sub-contractor stated therein but also of any particular conditions or require-ments of the contractor which he may wish to raise with the sub-contractor'.

Figures 5.4 and 5.5 show pages 9 and 10 of schedule 2 (it contains four pages in all) to illustrate how it is envisaged that they might be completed.

The architect's contribution is shown as typeface, that of the sub-contractor in longhand and the main contractor's insertions are in block letters.

Notes about which party is to complete which section are provided on page 2 of NSC/1. In Fig. 5.4 (p. 9 of NSC/1) it can be seen that the period initially

Schedule 2: Particular Conditions

Note: When the Contractor receives Tender NSC/1 together with the Architect/Supervising Officer's preliminary notice of nomination under clause 35·7·1 of the Main Contract Conditions then the Contractor has to settle and complete any of the particular conditions which remain to be completed in this Schedule in agreement with the proposed Sub-Contractor. The completed Schedule should take account not only of the preliminary indications of the Sub-Contractor stated therein, but also of any particular conditions or requirements of the Contractor which he may wish to raise with the Sub-Contractor.

1.A   Any stipulation as to the period/periods when Sub-Contract Works can be carried out on site:[s]

See 1C JS. AD

to be between ___1st June 1983___ and ___29th August 1983___

Period required by Architect to approve drawings after submission   ___4 Weeks___   *JS*

1.B   Preliminary programme details[t] (having regard to the information provided in the invitation to tender)

Periods required:

(1) for submission of all further sub-contractors drawings etc. *(co-ordination, installation, shop or builders' work, or other as appropriate)*[u]

~~not applicable~~ *AD JS*

~~not applicable~~ *AD JS*

2 weeks   AD

(2) for execution of Sub-Contract Works: off-site   *4 weeks   AD*

on-site   ~~*6 weeks*~~ *AD*

Notice required to commence work on site   *4 weeks   AD*

1.C   Agreed programme details (including sub-contract completion date: see also Sub Contract NSC/4, clause 11·1)[v]

CARCASSING
COMMENCE ON 18 MAY 1983
COMPLETE BY 6 JUNE 1983   *AD   JS*

SECOND FIX & 1ST TEST
COMMENCE ON 20 JULY 1983
COMPLETE BY 8 AUGUST 1983

FINAL TEST & LIVEN UP
DURING W/E 10 AUGUST 1983

---

2.   Order of Works to follow the requirements, if any, stated in Schedule 1, item 11[w]

Fig. 5.4 Conditions for agreement, NSC/1, page 9

72

3.A Attendance proposals (other than †general attendance).[×]

(a) Special scaffolding or scaffolding additional to the Contractor's standing scaffolding.

Boarded scaffolding for outside lights and ~~dining hall area~~   MOBIL TOWER WITH
        JS   *2m × 1.5m* PLATFORM IN
         DINING HALL AREA
                AD   JS

(b) The provision of temporary access roads and hardstandings in connection with structural steelwork, precast concrete components, piling, heavy items of plant and the like.

not applicable   AD

(c) Unloading, distributing, hoisting and placing in position giving in the case of significant items the weight and/or size. (To be at the risk of the Sub-Contractor).

none   AD

(d) The provision of covered storage and accommodation including lighting and power thereto.

Weatherproof and securable shed 4m × 3m
for fittings with light and power points   AD
METERED ELECTRICAL SUPPLY WILL BE
PROVIDED AND SUBCONTRACTOR WILL BE   JS
CHARGED FOR POWER CONSUMED AD

(e) Power supplies giving the maximum load.

for handtools only   AD

(f) Maintenance of specific temperature or humidity levels.

none   AD

(g) Any other attendance not included under (a) to (f) or as †general attendance under Sub-Contract NSC/4, paragraph 27·1·1.

none   AD

†Note: For general attendance see clause 27·1·1 of Sub-Contract NSC/4 which states: "General attendance shall be provided by the Contractor free of charge to the Sub-Contractor and shall be deemed to include only use of the Contractor's temporary roads, pavings and paths, standing scaffolding, standing power operated hoisting plant, the provision of temporary lighting and water supplies, clearing away rubbish, provision of space for the Sub-Contractor's own offices and for the storage of his plant and materials and the use of messrooms, sanitary accommodation and welfare materials." See SMM, 6 edn., B.9.2.

Fig. 5.5 Conditions for agreement, NSC/1, page 10

73

suggested by the architect in 1A as being when he envisaged the works will be carried out has been varied to some extent by the main contractor and electrical sub-contractor in 1C by mutual agreement. It may be that in the event contract progress will require the sub-contractor to bring forward or put back the periods when he will be able to work on the site. In 1B it can be seen that the parties have agreed that the sub-contractor will be given four weeks notice to commence on site.

In 3A (a) a compromise has been reached between main contractor and sub-contractor regarding the means of access, to the 'high' area of the dining-hall. The initials of both parties opposite the acceptable wording can be taken as confirmation of their respective intentions. Clearly, all the parties to the contract will find it necessary to think through the project to a considerable extent at pre-contract stage, if they are to complete NSC/1 in a meaningful manner. The aim of pre-contract negotiation is to prevent, or at least substantially to reduce, post-contract misunderstandings.

Employer/Nominated Sub-Contractor Agreement NSC/2

The form 'Agreement NSC/2' is returned, completed, by the sub-contractor at the same time as his tender contained in NSC/1.

NSC/2 extends to four pages, two-and-a-half of which are concerned with the *agreement clauses*. It is an amplified version of the warranty agreement between the employer and the nominated sub-contractor formerly issued by the RIBA for use with the 1963 edition of the Standard Form of Building Contract. NSC/1 comes into operation at two stages; clauses 1 and 2 operate after tender NSC/1 has been approved on behalf of the employer by signature on page 1, and the remaining clauses after the *nomination* of the proposed sub-contractor on form 'Nomination NSC/3'. Reference to the implications of stipulation 1 on page 2 of NSC/1 in this connection was made earlier in this chapter.

The first part of NSC/2 is the *agreement* between the employer and the sub-contractor; the sub-contractor is *not a nominated* sub-contractor at this stage.

Probably the most important point to ensure when completing page 1 is that the date to be inserted must be the date when tender NSC/1 form is signed as 'approved' by the architect on behalf of the employer. There is reference to this on page 1 of NSC/2 as can be seen in Fig. 5.6.

The four *recitals* on page 1 of NSC/2 (the 'whereases') are generally self-explanatory; they refer to the procedure as a result of which agreement NSC/2 is to be executed. The fourth recital is interesting in that it 'exempts', so to say, the architect from liability for the contents of the agreement and from liability for the contents of the tender.

NSC/2 conditions

A number of points require clarification in the conditions contained in NSC/2.

The actual wording of each clause is shown in small print and these clauses are followed by a brief statement by way of explanation.

Clause 1

1.1 The Sub-Contractor shall, after the Architect has issued his preliminary notice of nomination under clause 35.7.1 of the Main Contract Conditions, forthwith seek to settle with the Main Contractor the Particular Conditions in Schedule 2 of the Tender.

1.2 The Sub-Contractor shall, upon reaching agreement with the Main Contractor on the Particular Conditions in Schedule 2 of the Tender and after that Schedule is signed by or on behalf of the Sub-Contractor and the Main Contractor, immediately through the Main Contractor so inform the Architect . . .

This clause is concerned with the sub-contractor's obligations to follow the procedures (which the main contractor is obliged by SF 80, cl. 35 to follow), to reach agreement with the main contractor regarding schedule 2 of NSC/1.

Clause 2

2.1 The Sub-Contractor warrants that he has exercised and will exercise all reasonable skill and care in

   .1 the design of the Sub-Contract Works in so far as the Sub-Contract Works have been or will be designed by the Sub-Contractor; and

   .2 the selection of materials and goods for the Sub-Contract Works in so far as such materials and goods have been or will be selected by the Sub-Contractor; and

   .3 the satisfaction of any performance specification or requirement in so far as such performance specification or requirement is included or referred to in the description of the Sub-Contract Works included in or annexed to the Tender.

Nothing in clause 2.1 shall be construed so as to affect the obligations of the Sub-Contractor under Sub-Contract NSC/4 in regard to the supply under the Sub-Contract of workmanship, materials and goods.

2.2 .1 If, after the date of this Agreement and before the issue by the Architect of the instruction on Nomination NSC/3 under clause 35.10.2 of the Main Contract Conditions, the Architect instructs in writing that the Sub-Contractor should proceed with

   .1 the designing of, or

   .2 the proper ordering or fabrication of any materials or goods for the Sub-Contract Works the Sub-Contractor shall forthwith comply with the instruction and the Employer shall make payment for such compliance in accordance with clauses 2.2.2 to 2.2.4.

   .2 No payment referred to in clauses 2.2.3 and 2.2.4 shall be made after the issue of Nomination NSC/3 under clause 35.10.2 of the Main Contract Conditions except in respect of any design work properly carried out and/or materials or goods properly ordered or fabricated in compliance with an instruction under clause 2.2.1 but which are not used for the Sub-Contract Works by reason of some written decision against such use given by the Architect before the issue of Nomination NSC/3.

.3 The Employer shall pay the Sub-Contractor the amount of any expense reasonably and properly incurred by the Sub-Contractor in carrying out work in the designing of the Sub-Contract Works and upon such payment the Employer may use that work for the purposes of the Sub-Contract Works but not further or otherwise.

.4 The Employer shall pay the Sub-Contractor for any materials or goods properly ordered by the Sub-Contractor for the Sub-Contract Works and upon such payment any materials and goods so paid for shall become the property of the Employer.

.5 If any payment has been made by the Employer under clauses 2.2.3 and 2.2.4 and the Sub-Contractor is subsequently nominated in Nomination NSC/3 issued under clause 35.10.2 of the Main Contract Conditions to execute the Sub-Contract Works the Sub-Contractor shall allow to the Employer and the Main Contractor full credit for such payment in the discharge of the amount due in respect of the Sub-Contract Works.

Clause 2 is generally concerned with design, materials and performance specification.

Clause 2.1 sets out the sub-contractor's design warranty. The warranty comes into effect when the sub-contractor and employer enter into agreement NSC/2 and when the employer (or the architect on his behalf) approves tender NSC/1.

Not until this has been done should design works be carried out or material be ordered prior to nomination and as one of the objects of this arrangement is to provide a framework in which the design *may* be put in hand the importance of the sub-clause will be apparent.

Clause 2.2 makes provision for the sub-contractor to be paid for design work and for materials ordered if no nomination is made. In this event the employer may, upon payment, use the design work and become owner of the materials. If the sub-contractor is *not* prepared to let his design or materials be used by another contractor he should make this clear in his offer to the architect.

Clause 3

3.1 The Sub-Contractor will not be liable under clauses 3.2, 3.3 or 3.4 until the Architect has issued his instruction on Nomination NSC/3 under clause 35.10.2 of the Main Contract Conditions nor in respect of any revised period of time for delay in carrying out or completing the Sub-Contract Works which the Sub-Contractor has been granted under clause 11.2 of Sub-Contract NSC/4.

3.2 The Sub-Contractor shall so supply the Architect with such information (including drawings) in accordance with the agreed programme details or at such time as the Architect may reasonably require that the Architect will not be delayed in issuing necessary instructions or drawings under the Main Contract, for which delay the Main Contractor may have a valid claim to an extension of time for completion of the Main Contract Works by reason of the Relevant Event in clause 25.4.6 or a valid claim for direct loss and/or expense under clause 26.2.1 of the Main Contract Conditions.

3.3   The Sub-Contractor shall so perform his obligations under the Sub-Contract that the Architect will not by reason of any default by the Sub-Contractor be under a duty to issue an instruction to determine the employment of the Sub-Contractor under clause 35.24 of the Main Contract Conditions provided that any suspension by the Sub-Contractor of further execution of the Sub-Contract Works under clause 21.8 of Sub-Contract NSC/4 shall not be regarded as a 'default by the Sub-Contractor' as referred to in clause 3.3.

3.4   The Sub-Contractor shall so perform the Sub-Contract that the Contractor will not become entitled to an extension of time for completion of the Main Contract Works by reason of the Relevant Event in clause 25.4.7 of the Main Contract Conditions.

This clause deals with delay in supply of information and in performance by the sub-contractor.

It requires the sub-contractor to provide the architect with information (including drawings) in accordance with an agreed programme BUT (and it is a big 'but') this requirement only comes to pass when the architect has issued his instruction on 'Nomination' NSC/3, form.

Clause 4

4   The Architect shall operate the provisions of clause 35.13.1 of the Main Contract Conditions.

Clause 35 of the main contract is the 1980 edition version of clause 27 (nominated sub-contractors), in the 1963 edition of the Standard Form of Building Contract and sub-clause 35.13.1 is concerned with 'Payment of Nominated Sub-Contractor'.

Clause 35 is discussed in Chapter 16; sufficient to say for the time being that the NSC/2, clause 4, refers to the procedure by which the architect directs on value of sub-contract work in interim certificates and the information he shall give the sub-contractor which is, incidently, more comprehensive than it has been in the past.

Clause 4 provides the (by this time) *nominated* sub-contractor witn a direct right against the employer to ensure that the architect directs the main contractor as to the amounts for the sub-contractor which are included in the amounts stated as due in interim certificates. This reduces the sub-contractor's dependence upon the main contractor in this respect. This is a similar arrangement to that which exists in the 1963 conditions of contract.

Clause 5

5.1   The Architect shall operate the provisions in clauses 35.17 to 35.19 of the Main Contract Conditions.

5.2   After due discharge by the Contractor of a final payment under clause 35.17 of the Main Contract Conditions the Sub-Contractor shall rectify at his own cost (or if he fails so to rectify, shall be liable to the Employer for the costs referred to in clause 35.18 of the Main Contract Conditions) any omission, fault or defect in the Sub-Contract Works which the Sub-Contractor is bound to rectify under Sub-Contract NSC/4 after written notification thereof by the Architect at any time before the issue of the Final Certificate under clause 30.8 of the Main Contract Conditions.

5.3   After the issue of the Final Certificate under the Main Contract Conditions the Sub-Contractor shall in addition to such other responsibilities, if any, as he has under this Agreement, have the like responsibility to the Main Contractor and to the Employer for the Sub-Contract Works as the Main Contractor has to the Employer under the terms of the Main Contract relating to the obligations of the Contractor after the issue of the Final Certificate.

Just as clause 4 was concerned with *interim* payments, clause 5 is concerned with *final* payments. It *provides the employer with a duty to make final payment to the sub-contractor* but it also sets out *obligations on the sub-contractor's part*.

It provides the sub-contractor with a direct right against the employer to ensure that the employer follows the appropriate procedures, set out in clause 35.17 of SF 80, regarding early final payment in return for which the sub-contractor is obliged *direct* to the employer to rectify faults in the sub-contract works before and after issue of the final certificate under the main contract. Clause 5 should be read in conjunction with clause 35.18 of SF 80.

Clause 6

6   Where the Architect, has been under a duty under clause 35.24 of the Main Contract Conditions except as a result of the operation of clause 35.24.6 to issue an instruction to the Main Contractor making a further nomination in respect of the Sub-Contract Works, the Sub-Contractor shall indemnify the Employer against any direct loss and/or expense resulting from the exercise by the Architect of that duty.

Clause 6 places upon the architect a duty to make a further nomination in the event of the sub-contractor making default in respect of a number of matters referred to in clause 29 of NSC/4 in the event of him becoming bankrupt or determining his employment under clause 30 of the same form.

Clause 6 gives the employer a *direct right against the nominated sub-contractor* for an indemnity against the direct loss and/or expense which the employer may incur if the architect, because of some fault, etc. of the sub-contractor, has had to issue a further nomination to carry out and complete the sub-contract works.

Clause 7

7.1   The Architect and the Employer shall operate the provisions in regard to the payment of the Sub-Contractor in clause 35.13 of the Main Contract Conditions.

7.2   If, after paying any amount to the Sub-Contractor under clause 35.13.5.3 of the Main Contract Conditions, the Employer produces reasonable proof that there was in existence at the time of such payment a petition or resolution to which clause 35.13.5.4.4 of the Main Contract Conditions refers, the Sub-Contractor shall repay on demand such amount.

In the event of the main contractor *not* making payment to the sub-contractor (not an unknown situation), clause 7 provides for the architect to issue a certificate on non-discharge by the main contractor thus providing for an alternative means of payment.

This clause *provides for direct payment of the sub-contractor by the employer* under clause 35.13 of the main contract. This he, the employer, would pay for out of monies which he has by reducing the amounts otherwise due to the main contractor.

There is an interesting provision for *repayment* to the employer by the sub-contractor in clause 7.2 if there is in existence 'either a petition which has been presented to the Court or a resolution properly passed for the winding up of the main contractor other than for the purposes of amalgamation or reconstruction, whichever shall have first occurred'.

Clause 8

8   Where clause 2.3 of Sub-Contract NSC/4 applies, the Sub-Contractor shall forthwith supply to the Contractor details of any restriction, limitation or exclusion to which that clause refers as soon as such details are known to the Sub-Contractor.

Clause 2.3 of sub-contract NSC/4 is concerned with restrictions in contracts of sale, etc. and it *limits the liability of the sub-contractor*. It applies if the sub-contractor is required to enter into a sub-sub-contract or a contract of sale with a person or persons other than the main contractor. It will, however, be seen in clause 8, above, that the sub-contractor is required to supply to the main contractor details of any restriction, limitation or exclusion to which condition 2.3 of NSC/4 refers as soon as he, the sub-contractor, knows the details.

Clause 9

9   If any conflict appears between the terms of the Tender and this Agreement, the terms of this Agreement shall prevail.

This clause requires no amplification.

Clause 10

Clause 10 is concerned with arbitration. This subject is discussed in Chapter 14.

Form of nomination – NSC/3

'JCT Standard Form of Nomination of Sub-Contractor where Tender NSC/1 has been used' (to give it its full title), is reproduced as Fig. 5.6. Nomination NSC/3 is issued by the architect under SF 80 clause 35.10.2 and upon its issue it is

JCT

JCT Standard Form of Nomination of Sub-Contractor where Tender NSC/1 has been used

To: P. ROPER Builders Ltd.

64 High Street, Worsten, BEDS.
(Main Contractor)

Main Contract Works: Community Centre for Worsten District Council

Job reference: CCW/29/82

Sub-Contract Works: Electrical Installation

Page Number – Bills of
Quantities or Specification: Specification pp. 90-96 incl.

Sub-Contractor hereby
nominated: J. Shine Electrical Services Ltd.

of: 10 High Street, Nearbigh, BEDS.

Tel. No: 0123-45678

Further to my/our Preliminary Notice of Nomination (Main Contract Conditions clause 35·7·2)

dated 1st September   19 82

and Tender NSC/1 (and annexed documents) duly completed by you and the Sub-Contractor named above and by myself/ourselves on behalf of the Employer, the Sub-Contractor named above is hereby **NOMINATED** under the Main Contract Conditions clause 35·10·2 for the Sub-Contract Works identified above.

Signed _____ A. D. Signer RIBA _____ Architect/Supervising Officer

Address 64 High Street, Worsten, BEDS.

Date 19th September   19 82

Circulation
☐ Main Contractor   ☐ Clerk of Works
☐ Quantity Surveyor   ☐ Consulting Engineer
☐ Sub-Contractor hereby nominated   ☐ Architect/Supervising Officer's file

© 1980 RIBA Publications Ltd

Fig. 5.6 Form of Nomination, NSC/3

80

intended that the nominated sub-contract should come into existence. Reference to Fig. 5.3 shows a copy of page 1 of NSC/1 at the foot of which it can be seen that when the main contractor signs NSC/1 he does so 'subject to a nomination instruction on Nomination NSC/3 under clause 35.10 of the Main Contract Conditions'.

The main contractor and the nominated sub-contractor then enter into sub-contract NSC/4.

This concludes the series of 'operations' shown in the network in Fig. 5.2 which comprise the procedure using the basic method. It is intended that the basic method be used for the vast majority of contracts and especially where major specialist works are concerned or where the sub-contractor makes a significant contribution to the design of the works.

The alternative method

When should the alternative method be used?

There is no doubt that the Joints Contracts Tribunal consider that the alternative method should be used as little as possible. It has been suggested above that one of the main aims of the basic method in the new sub contract procedures is to attempt to settle possible matters of dispute between the main contractor and the sub-contractor *before* they arise, indeed, before the parties have even entered into a contract. It is what the medical profession would call preventative medicine.

It follows, then, that as the alternative method leaves numerous matters to be resolved *after* the sub-contractor has been nominated, it is not to be encouraged. The JCT *Guide* says (p. 63): 'It is suggested that the basic method may be found to be the more convenient arrangement for nominated sub-contract tenders which can be approved by the Employer *before* the Main Contract is let. The preferences of prospective Nominated Sub-Contractors should always be an important consideration in deciding which method to adopt. The alternative method may be considered suitable particularly where the Nominated Sub-Contract Works may not be critical to the progress of the Works as a whole. However, even in these circumstances where the Contractor's programme will have been settled . . . since the scope for a proposed sub-contractor on commencement dates and periods on site, etc. is inevitably limited by the Contractor's programme. It may well be found that . . . the basic method would be appropriate . . . ' The wording regarding the 'preferences of prospective Nominated Sub-Contractors' is of significance for specialist contractors. Building and installation works which are an integral part of most buildings are almost invariably 'critical to the progress of the Works as a whole'.

There is little doubt but that the basic method of nomination should be used whenever building services, electrical and similar sub-contractors are employed. Indeed, apart from window-box planters and flag-pole installers it is difficult to think of contractors who could not, with a modest stretch of the imagination, be regarded as 'critical' on some contracts and perhaps if a royal opening ceremony were envisaged even gaily planted window-boxes and operative flag-poles might appear on the critical path!

Using the alternative method

How is the alternative method used? The most important clauses in the *main contract* read:

35.5 .1 .1 Tender NSC/1 and Agreement NSC/2 shall be used in respect of any part of the Works for which a sub-contractor will be nominated by the Architect unless clause 35.5.1.2 is operated.

.1 .2 The Contract Bills, or any instruction under clause 13.2 requiring a Variation (including an instruction under clause 35.5.2) or under clause 13.3 on the expenditure of a provisional sum, may state that clauses 35.11 and 35.12 (Tender NSC/1 and Agreement NSC/2 not used) shall apply to any part of the Works for which a sub-contractor will be nominated by the Architect and where so stated the Contract Bills or the instruction shall also state whether or not the proposed sub-contractor has tendered, or has been or will be asked to tender on the basis that Agreement NSC/2a will be used.

Tender NSC/1 and Agreement NSC/2 not used
35.11 Where clause 35.5.1.2 has been operated:

35.11 .1 the Employer shall enter into Agreement NSC/2a with the proposed sub-contractor (unless the tender of the proposed sub-contractor referred to in clause 35.5.1.2 has been requested and submitted and approved on behalf of the Employer on the basis that Agreement NSC/2a shall not be entered into) and

35.11 .2 the Architect shall issue an instruction to the Contractor (with a copy to the proposed sub-contractor) nominating the proposed sub-contractor to supply and fix the materials or goods or to execute the work.

35.12 The Contractor shall proceed so as to conclude a sub-contract on Sub-Contract NSC/4a with the proposed sub-contractor within 14 days of the nomination instruction under clause 35.11.

Once again, the procedures can best be understood by using a diagram such as that shown in Fig. 5.7.

From the above clauses and from the diagram it can be seen that any instruction of the architect may state that the alternative method shall be used and that if such a statement is made then the contract bills *or* the architect's instruction shall also state 'whether or not the proposed sub-contractor has tendered, or has been or will be asked to tender on the basis that Agreement NSC/2a will be used'.

Clause 35.11 also provides for the use of agreement NSC/2a *unless* the proposed sub-contractor has tendered and his tender has been approved on behalf of the employer on the basis that agreement NSC/2a shall *not* be entered into.

Nodes 4A and 4B in Fig. 5.7 mark the point at which the architect must decide

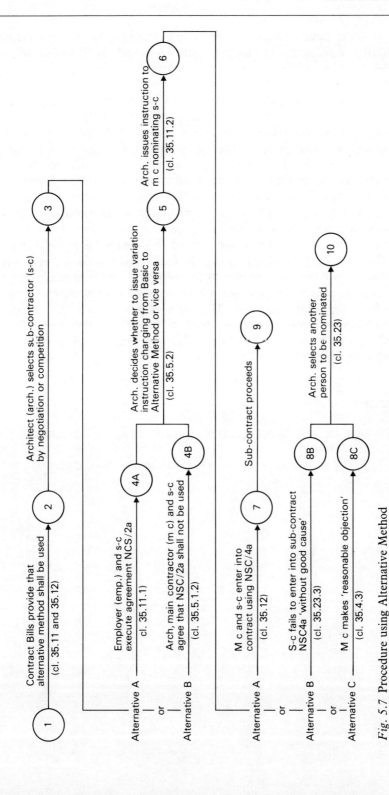

Fig. 5.7 Procedure using Alternative Method

whether or not to issue a variation instruction changing from the basic method to the alternative method – or vice versa (op. 4A–5 and 4B–5) and in op. 5–6 the architect issues instructions to the main contractor nominating the sub-contractor.

At this point alternatives again appear. Only the first of these alternatives (op. 6–7), leads to a sub-contract using sub-contract NSC/4a.

Alternatives B and C (ops 6–8B and 6–8C) provide the main contractor and the sub-contractor with 'escape routes' if, in the case of the sub-contractor, 'without good cause [he] fails within a reasonable time to enter into the sub-contract NSC/4a', or in the case of the main contractor, he makes 'any such reasonable objection at the earliest practicable moment but in any case not later than 7 days from receipt by him of the instruction of the architect under clause 35.11 nominating the sub-contractor'.

In the event of either the situations referred to in ops 6–8B or 6–8C it is up to the architect to select another person to be nominated (ops 8B–10 or 8C–10). Sub-contractors will find that the alternative method is similar in many respects to the procedure used for nomination where the 1963 edition of the Standard Form of Building Contract is used. This in itself may suggest that it will be looked upon kindly by some (better the devil you know!) but with the familiarity of procedures may go the familiar crises which have been experienced with the 1963 method of nomination.

References

Banwell Committee (1964), *The Placing and Management of Contracts for Building and Civil Engineering Work*. HMSO.
Salzman, L. F. (1967) *Building in England*. Oxford University Press, Oxford.

Chapter 6

Entering into a contract: obligations and action to take

(The word 'contractor', where used in this chapter, applies to specialist (sub-) contractors)

This chapter is concerned with aspects of law in so far as they affect the sub-contractor. The aim is to point out to the sub-contractor some of the pitfalls of which he should be aware. It has been said, perhaps by a lawyer, that a man who acts as his own lawyer has a fool as a client. Certainly, if a contractor has doubts about his position in law he should consult an expert adviser in the form of his own solicitor of, perhaps, the legal department of his trade federation. But a contractor cannot have a lawyer perpetually in his pocket and as John Selden said more than 300 years ago 'Ignorance of the law excuses no man' so an elementary appreciation of the law may save the contractor money and worry.

As far as contracts are concerned, most contractors, wisely, think in terms of standard forms of contract, but it is not in law necessary for him to sign a contract in order to make a legally binding agreement. It is possible for him to enter into a contract without realizing it. What, then, are the fundamental features of a contract which are likely to affect the contractor.

A contract may be defined as an agreement between parties which contemplates and creates an obligation.

The essentials in the formation of a contract are shown in Fig. 6.1. They are:
(a)   agreement between the parties;
(b)   an offer;
(c)   an aceptance of the offer;
(d)   a consideration (the price to be paid for the service or goods to be provided).

Agreement offer and acceptance

An agreement between the parties regarding the purpose, rights and the obligations which the contract will create is essential. In order to create an obligation the agreement must:
(a)   be for a lawful purpose;
(b)   be between parties who have the capacity to enter into a contract;
(c)   be free from misrepresentation, mistake, duress or other circumstances which will vitiate (invalidate) it.

Contractors should ensure that they do not enter into a contract which contra-

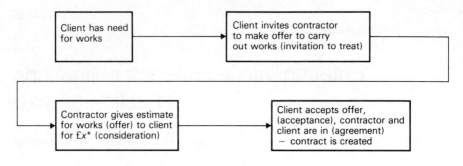

(Wording in brackets) indicates fundamental requirements
of a contract

*the sum need not be stated but agreement
must provide for payment, in some form,
for the works which the contractor is to carry out

Fig. 6.1 Stages in the creation of a contract between client and contractor

venes the Building Regulations or similar, Town Planning Acts or other Acts. Infants or minors (persons under 18 years of age), persons of unsound mind and drunkards are incapable of entering into contracts for most if not all of the services which a contractor would be likely to provide so he should be wary of dealing with them. He should be wary, too, of entering into agreements with officers and councillors of local authorities and ensure that contracts with local authorities and 'corporations' are under the seal of the local authority or corporation. Local authorities and most corporations have the interests of the contractor as well as their own at heart when they enter into contracts but contractors may be well advised to take legal advice when contemplating a contract with an unknown corporate body.

Contractors' directors who are involved in local affairs – councillors and the like – should be wary of entering into contracts with their own council, etc.

Offer and acceptance

A contract is effected when an offer made by the offerer is accepted by the offeree. An offer may be in writing or it may be oral. It may even be in the form of a gesture.

An essential aspect of an offer is that the offerer intends to be bound by the terms if the offer is accepted. In this respect an offer is different from a declaration of intention and a statement of price which is known as an 'invitation to treat'.

An example of an invitation to treat which is familar to most contractors is an auction sale. When the auctioneer calls up the lot he is inviting those present to treat as his 'what am I bid for this. . .' indicates. The persons interested in the lot make their respective bids (offers) and the auctioneer binds himself to accept the offer when he bangs his hammer. An example of a gesture or sign being sufficient to create a contract may be seen at an auction where the bidder frequently indicates his intention by a nod, wave of catalogue or other means.

An invitation from a prospective client or builder to a contractor to 'give him a price', (tender for work), is another example of an invitation to treat. If the contractor offers to carry out the work for £x and the client accepts the contractor's offer, an agreement has been made and a contract has been created. Both (or either) offer and acceptance may be oral or in writing – the contract will be binding on the parties concerned.

To summarize so far, in general terms a contract is created:

(a)   provided the agreement is not for works which contravene Building Regulations or other regulations or some Act of Parliament (is not illegal);

(b)   provided the parties are not of unsound mind, minors or, if limited liability companies, the purpose of the contract falls within the 'objects' clause of the company's memoranda of association;

(c)   provided a clear offer is given an unconditional acceptance (neither offer nor acceptance need be in writing);

(d)   provided one party obtains an advantage (receives payment) or detriment is suffered by the other in exchange (makes payment) for a promise (the consideration).

Oral contracts

As has been mentioned above, these are binding on both parties and they are frequently quite satisfactory until or unless one or the other parties finds things are not going as he anticipated. The aggrieved party then finds that there is no written evidence of the parties' intentions.

Contractors occasionally run into difficulties in this way when they start work on a project on the understanding that a written contract, frequently using a standard form of contract, will be entered into in due course. Once the contractor has commenced work he is in a vulnerable position which usually becomes more vulnerable as the work progresses. The contractor may find the client (or builder) much less interested in signing a standard form of contract than he would have been initially.

Why should he? He is almost certainly stronger as things are – particularly if he normally uses (or even placed an order using) his own order form with conditions printed on the back which were drafted by his solicitor to give him the best possible terms. As things stand, all the conditions necessary to create a contract have been created – offer, acceptance, consideration, etc. but there may be no evidence of, say, the contractor's expectation of payments on account as the works progress. The contractor's willingness to respond to the client's (or builder's) request to 'get things started to help the job along' could have unfortunate repercussions and the contractor may discover the wisdom of the quotation attributed to Sam Goldwyn, the Hollywood film magnate, that 'an oral contract is not worth the paper it's written on'. A simple exchange of letters, the first making an offer the second accepting it, will frequently prevent future misunderstandings.

Think before signing and sign before starting. The contractor's golden rule, should be to take care to read the conditions of a contract before he signs any agreement and to sign, and see that the other party signs, before starting work or incurring any cost or expense by, say, ordering materials for use on the project. It must be remembered that the personnel in companies change quite frequently. During the life of a project it is possible that the people with whom the contractor started so amicably will move on leaving the contractor dealing with less amenable people. It is an imprudent contractor who does not ensure at the outset that his contractual position is secure lest, perhaps in two or three years time, an atmosphere between the partners concerned which was all sweetness and light during the heady pre-contract days turns sour and/or the light fades. The contractor may find it appropriate to remind his opposite number of the transitory nature of staff employment if it is inferred that the contractor is being less than fully cooperative by insisting on tying up all the 'paper work' before making a start.

Express and implied terms

Express terms are those statements or promises which are made by the parties to the contract. An undertaking by the contractor that he will complete the works by a certain date, or by the client that he will make payments on account when certain stages of construction have been reached, are examples of express terms.

Implied terms may arise by custom or trade practice, and by statute or legal precedents established in the courts from cases which have been contested in the past.

It is, for example, implied that goods or services to be provided will be fit for their purpose. An express term will normally take precedence over an implied term if, for example, the express term contradicts a normal trade custom but an express term must not be contrary to an obligation implied by law.

Privity of contract

Only the parties to a contract may normally benefit from or be bound by the terms of a contract. A third party cannot normally sue upon a contract even if the contract has been made expressly for the third party's benefit.

Assignment

The general rule is that the parties to a contract may not assign their benefits or obligations to a third party without the consent of the party with whom they have entered into a contract. There are exceptions to this general rule but most standard forms of contract require agreement between the contracting parties as a condition of assignment.

The foregoing notes provide only an introduction to an extensive area of study *from the standpoint of a contractor*. Contractors or students wishing to study contract law in detail will find the books listed at the end of this chapter of value.

Sub-contractor's obligations

What in specific terms is the sub-contractor obliged to do?

Primary obligation

Nominated sub-contract NSC/4 takes the place of the old green form and sets out the primary obligation of the sub-contractor in clause 4.1.1 as 'to carry out and complete the Sub-Contract Works'. In full it reads:

4.1 .1 The Sub-Contractor shall carry out and complete the Sub-Contract Works in compliance with the Sub-Contract Documents using materials and workmanship of the quality and standards therein specified and in conformity with all the reasonable directions and requirements of the Contractor (so far as they may apply) regulating for the time being the due carrying out of the Main Contract Works provided that where and to the extent that approval of the quality of materials or of the standards of workmanship is a matter for the opinion of the Architect, such quality and standards shall be to the reasonable satisfaction of the Architect.

This clause is very similar to the main contractor's obligation contained in clause 2.1 of the main contract and it is important to stress that the sub-contractor should be familiar with the main contract conditions.

Secondary obligations

Conform with contractor's directions
There have been only minor drafting changes from the wording contained in the 1963 edition as far as the main contract is concerned. Clause 4.1.1 in NSC/ 4 has been brought more in line with the wording of the main contract than was the case with the old green form.

The principal difference between the obligation of the main contractor and the sub-contractor is that the sub-contractor has to conform with all the *reasonable directions and requirements* of the main contractor in addition *to satisfying the architect* with regard to the quality of materials and the standards of workmanship.

There are further obligations in clause 4 with regard to
(a) the sub-contractor keeping a 'competent person-in-charge' upon the sub-contract works;
(b) regarding the issue of instructions by the architect and with which the sub-contractor 'shall forthwith comply' but these matters will be left for the time being in order to concentrate on clause 5.1.1 in NSC/4 which sets out the

sub-contractor's liability 'under incorporated provisions of the Main Contract'. Clause 5.1.1 reads:

5.1 The Sub-Contractor shall:

5.1 .1 observe, perform and comply with all the provisions of the Main Contract as described by or referred to in the Tender, Schedule 1 on the part of the Contractor to be observed, performed and complied with so far as they relate and apply to the Sub-Contract Works (or any portion of the same). Without prejudice to the generality of the foregoing, the Sub-Contractor shall observe, perform and comply with the following provisions of the Main Contract Conditions: clauses 6, 7, 9, 16, 32, 33 and 34:

If the sub-contractor is to comply with all the provisions of the main contract so far as they relate and apply to the sub-contract works he must have access to a copy of the main contract so that he is aware of any deletions from, or additions to, the conditions which may have been agreed between the employer (client) and the main contractor. The clauses in the main contract to which particular reference is made in clause 5.1.1 of NSC/4 refer to:

Cl. 6 Injury to persons and property – indemnity to main contractor
Cl. 7 Levels and setting out of the works
Cl. 9 Royalties and patent rights
Cl. 16 Materials and goods unfixed or off-site
Cl. 32 Outbreak of hostilities
Cl. 33 War damage
Cl. 34 Antiquities

The sub-contractor should, however, bear in mind that these clauses have been mentioned 'without prejudice to the generality of' the other provisions made in clause 5.1.1 of NSC/4. There really is no alternative open to the sub-contractor other than to obtain a copy of the Standard Form in addition to the forms which more directly concern him. The 'Tender, Schedule 1' referred to in clause 5.1.1 above, is tender NSC/1 which is discussed in Chapter 5.

The necessity for the sub-contractor to take care to complete schedule 1 at the time of tender, to which considerable importance was attached in that chapter, is now apparent.

The remainder of clause 5 is concerned with the sub-contractor's liability to 'indemnify and save harmless' the main contractor against and from

(a) any breach, non-observance or non-performance by the sub-contractor from any of the provisions referred to in clause 5.1.1;

(b) any act or omission of the sub-contractor which involves the main contractor in any liability to the employer under the same provisions.

There is a safeguard for the sub-contractor in clause 5.2 that nothing contained in the sub-contract documents shall be construed so as to impose any liability on the sub-contractor in respect of any act or omission or default on the part of the employer, the main contractor and his other sub-contractors nor create any privity of contract between the sub-contractor and the employer or any other sub-contractor.

Any agreement which the sub-contractor and employer may have as a result of them using Form of Employer/Nominated Sub-contractor Agreement, Agreement NSC/2, is excluded from clause 5.2

The term 'privity' (of contract) simply means 'any relation between two parties that is recognized by law' (OED).

What is really being said in clauses 4.1.1 and 5.1.1 of NSC/4 is that the sub-contractor has to behave like a competent workman and *get on with the job* (carry out and complete the sub-contract works).

He is well advised to leave matters of design to the designers – the architect and other 'consultants' – but if he finds that his business makes it necessary for him to become involved in design probably his best course is to form a separate design unit, apart from his contracting organization, which does *only* design work, and charges a separate fee for doing so. Alternatively, if he is called upon to design a system he might consider passing the work to a consultant. This way there can be no confusion about contractual responsibility *vis-à-vis* the architect, consulting engineers and sub-contractor should problems develop. Experience shows that whenever the sub-contractor *is* both designer and contractor trouble is, almost certain to arise sooner or later – usually sooner.

One advantage of the 1980 edition of the Standard Form of Building Contract and the associated documentation is that there is less likelihood of the sub-contractor entering into a contract without thinking through his obligations. Nevertheless, there are occasions when a main contractor may press him to 'make a start before the paperwork is tied up', perhaps to make amends for delay on the part of the architect in nominating the sub-contractor.

It is an imprudent sub-contractor who yields to such pressure. Once a sub-contractor (or any contractor, for that matter) has commenced work without reaching agreement on all matters his contractual position is weaker than it would otherwise be because he has, to a considerable extent, forfeited his negotiating strength.

Domestic sub-contracts – action to take

So far this chapter has been concerned with contracts in general and with nominated sub-contracts in particular. It is by no means unknown for the specialist sub-contractor to tender for the specialist works content of a building project when the architect has *not* made this work the subject of a prime cost sum but has left it to the main contractor to deal with. It is then up to the main contractor to submit a tender and if he has a specialist contracting subsidiary within his group of companies he will probably pass a copy of the specialist works section of the bills or other information to that subsidiary for pricing.

However, not all main contractors have such specialist subsidiaries and they invite tenders from two or three contractors in the same way as they invite tenders for much of the other work.

What, then, should the sub-contractor do to safeguard his interests when tendering with a view to becoming a domestic sub-contractor?

The conditions of contract contained in DOM/1, the standard form of contract for domestic sub-contracts, are essentially the same as those in NSC/4 so far as the appendices are concerned and the sub-contractor should endeavour to see that matters which would be 'negotiated' when using the Basic Method described in Chapter 5 are given similar attention before contracts using DOM/1 are signed. In this way he should be reasonably satisfied that misunderstandings regarding the attendances to be provided by the contractor, etc. are avoided. The sub-contractor should also be sure that he is aware of the contractor's expectation of him!

In Chapter 5 we considered tender and nomination procedures where the *basic method* of nomination is used and it is appropriate here to consider steps which should be taken by the sub-contractor when he is at the point of entering into a contract such as that provided by DOM/1 or similar forms which do not demand such a systematic approach as that demanded by the basic method.

The prospect of work

Let us assume that having submitted a tender, time has passed and there are indications that the tender is likely to be accepted and that a sub-contractor will in due course be asked to enter into a contract. This is when the foundations laid when preparing the *estimate* (the technical assessment of the project) which was converted into the *tender* (which took into account the commercial factors – see Ch. 4) will prove their worth. If a good job was done at tender stage it should not be too difficult to enter into a contract but sub-contractors frequently tend to lose confidence at this time.

As the lowest tenderer he begins to wonder how many mistakes have been made and whether he should withdraw his tender.

This change of standpoint is a factor which cannot be ignored. Estimators, like contractors (if they are not one and the same person) have to be born optimists or they would not retain their sanity and would certainly not get any work. Once a job is in prospect all that changes: the doubting Thomases question the credibility of the tender, the accountant questions the firm's ability to finance the project; and the contracts manager is confident that he does not have the manpower to carry it out. It is at this time that the leader must come forward, restore confidence, reassure doubters, rally the bank manager's support and get things teed up generally. There is never enough time when estimating to take as much care as one should. Now that there is the prospect of a contract the balance can be redressed to some extent at least.

Re-examine the tender

There will no doubt be pressure from the main contractor to sign the contract and whilst it would be a foolish sub-contractor who ignored a possible source of future business, he should avoid being over-pressurized and go through his papers carefully.

It might well be years rather than months before a sub-contractor hears from one source or another that the tender he submitted is being considered. A sub-contractor could be forgiven for assuming that he had heard the last of that project and thrown the papers away. There is an obvious lesson to learn from this.

If the papers have not been lost the sub-contractor should resurrect them and follow a similar procedure regarding his information requirement as that set out in the Basic Method of tendering which is discussed in Chapter 5 or, for smaller and/or less complex projects, work his way through the check-list described, later in the chapter.

Typical items for consideration are given below.

Time for commencement and completion
At the time of tender the sub-contractor may be required to begin the works at a time which the contractor may reasonably require, but not less than x weeks from the date of the written acceptance of his tender, and to complete the whole of the works within y weeks. The sub-contractor will have given undertakings in this respect.

Were the undertakings he gave realistic?

How do they measure up with what he is now required to undertake?

Now is the time to unfold the time-based programme he prepared as the basis of his estimate and look at it in more detail. If he did not actually prepare a programme at that time he should certainly do so now. Planning has been described as 'building on paper'. We will go into the methodology of planning and the importance of method statements in preparing a plan in Chapter 7 but for the time being a simple bar-chart worked out from the number of man-hours allowed for each operation should be sufficient. The discipline of planning makes one consider which items of work the main contractor must have carried out before the sub-contractor can make his contribution. He has to think of *dependency* and *sequence*. For example, one cannot run conduits for the ceiling light-fittings in a suspended ceiling until the ceiling carcase has been fixed and one cannot fix the fittings until the ceiling tiles (or whatever the finish may be) have been fixed. One should have thought the job through to that extent at least on paper before committing oneself to firm dates on a form of contract.

In order to prepare a programme to the extent referred to above it will almost certainly be necessary for the sub-contractor to refer to the contractor if he has not already referred to him. It makes for a more orderly job if the main contractor calls the sub-contractors together before the sub-contracts are signed and agrees the *strategic plan* with them. In this event the managers of main and sub-contractors can agree the *order of work* which it is up to the main contractor to determine.

The sub-contractor may avoid problems at a later date if he, at least, agrees the broad programme before he signs the contract.

Are adequate drawings available?
The sub-contractor is well advised to see that the drawings which are to be used for the project bear a passing resemblance to those on which he based his tender.

The first place to look for this information is in the Articles of Agreement of the main contractor's JCT form. In all probability the drawings will not actually be listed there because there is too little space but the articles will probably refer to an appendix to the bills of quantities, or some similar appendix, where a full list is available. Unless it is a small project the sub-contractor may have some cause for apprehension if it *is* possible actually to list the drawings in the articles – if it is, there are probably too few for comfort. Wherever they actually appear, the drawings referred to in the articles are the *contract drawings* and as such should be those on which the main contractor's contract is based.

The sub-contractor should find at least some of the drawings on which he based his tender listed in the articles. He should note the revision letters on the drawings as well as the drawing numbers themselves.

His next source of drawings is the list which should appear in Part 2 of the appendix to DOM/1 or other appendix if DOM/1 is not used. He will obtain a good indication of the changes which have occurred on the project by comparing the revision letters on the main contractor's articles list and the list in his own first recital.

If the drawing numbers differ greatly he will need to take a much closer look.

Watch for revised drawings
The object of this examination is to ascertain the extent to which the project has changed since he prepared his estimate. He should note the numbers and revision letters of all the drawings which were available when he visited the architect's or contractor's office. If he did this he has a datum from which to measure changes which have occurred in the meantime. He is not looking for difficulties: he is looking to avoid them later on. Not all revisions to the drawings will affect the sub-contract work; frequently the revisions are quite trivial and irrelevant as far as he is concerned. Any competent firm of architects or consulting engineers will have a column on each drawing in which to enter each revision.

Revision	Date	Description
A	20.5.78	Window fanlights revised
B	1.6.78	Window cill tiles changed to Eternit
C	7.6.78	Suspended ceiling omitted. Plaster and emulsion paint added.
D	12.7.78	Doors 1/1, 1/5, 1/9 hung opposite hand.

A glance at the revision column should enable the sub-contractor to see which revisions are likely to affect him. Occasionally, however, revisions are made on drawings which are *not* recorded so it is as well to scan the drawing. He should, in any event, have kept the drawings on which he based his tender (clearly marked 'TENDER DRAWING' in the bottom right-hand corner with his date stamp added) so he has a record of the information which was available if he needs it at a later date.

In the case of revisions A and B, above, the work is unlikely to be affected. Revision C could, however, materially affect his work. The conduits serving light fittings will now be placed above or in the structure of the floor rather than below it. This could affect his programme by making it necessary for him to advance his commencement dates affect the cost of labour and materials by varying them. Having recently prepared a programme of work based on his tender data, his memory will be fresh and he will be able to identify changes quite easily.

Revision D might or might not affect his work depending on the state of advancement of the project. If work is not far advanced it probably makes little difference if, say, switch-drops are on one side of a door or another.

The important thing for the sub-contractor is that the drawings listed in the contract documents do not represent more work than those on which he tendered. If they do he should draw the attention of the main contractor to them so that the cost of the variations can be adjusted from his tender to ensure that the contract sum incorporates the variations. Obviously quite a lot depends on the wording of the contract but if it reads that the sub-contractor undertakes to 'execute and complete the sub-contract works in accordance with the drawings listed below, etc.' and the listed drawings are different from, or revisions of, those on which the sub-contractor based his estimate, there could be trouble ahead.

Architects sometimes require sub-contractors to return the consultant's drawings with their tenders. This is an unreasonable practice which should not be advocated because trunking sub-contractors will almost certainly have marked their conduit runs, etc. on them. If, however, the architect ask for drawings to be returned the sub-contractor's courses of action are probably: (a) to take copies, if this is possible; (b) to overlook the instruction which is unlikely to be followed up; or (c) to make an appropriate 'stalling' reply.

Compare drawings

What should the sub-contractor look for when comparing the tender and sub-contract drawings? Assuming the sub-contractor is, say, a services engineer or similar specialist contractor, the answer is items such as:

(a) variations in the location of his intake;

(b) that the accommodation available for mains, trunking, elements, meters etc. has not varied;

(c) that the material to which the sub-contractor's trunking, conduits and fittings are to be fixed has not changed so that fixings will be more difficult or costly;

(d) that holes through walls, floors, etc. are where the sub-contractor planned for them to be or, if repositioned, that his runs are not greater than allowed for in his estimate;

(e) that applied finishes to ceilings, walls and floors, panelling, ductwork, tiling, etc. have not changed to the extent that they involve the sub-contractor in greater work or costs. (The change could be such that the task of fixing to the surface may be greater or such that the sequence of working will have to be changed to the detriment of the sub-contractor);

(f)   that the scope of his work has not increased.

Specification and/or bill of quantities

Much has been said, above, about the need to take care to check that the proposed sub-contract drawings do not vary from those on which the sub-contractor based his tender but no mention has so far been made of the specification or bills of quantities which will form part of the sub-contract documents. The reason for this is that in practice these do not pose such a problem for the sub-contractor. It is usually a simple matter to check the proposed specification/bill of quantities against the tender copy – they are either the same or they are not! If they are not the same the differences are easily identified and the tender price can be amended accordingly.

Inability to secure materials and/or labour

SF 80 provides for the main contractor to be given an extension of time if by his inability for reasons beyond his control and which he could not reasonably have foreseen at the date of his contract he is unable to secure such labour and/or such goods and/or materials as are essential to the proper carrying out of the works.

It is by no means unusual for the architect to delete either or both these sub-conditions from the main contractor's contract. If he does so the main contractor is unlikely to be prepared to give the sub-contractor an extension of time.

The sub-contractor should have decided at tender stage whether or not he would accept deletion of these sub-conditions. Obviously, commercial factors (such as needing the work!) will have guided his decision but sub-contractors who are subject to non-delivery of their materials or labour should consider carefully before agreeing to the deletion of conditions which might increase their liability.

Defects liability period

It should be remembered that the period referred to in the appendix to the main contract is the main contractor's and that his contract period may extend well beyond that of the sub-contractor. The architect will often agree to the sub-contractor's defects liability period commencing in advance of that of the main contractor if the sub-contractor's work is completed in advance of the main contractor's work, but it is not possible to resolve such questions at this stage. It is probably advisable for the sub-contractor to leave the question of the defects liability period and running parts of his installation and equipment in advance of the sub-contractor's overall completion until the question arises towards the end of the project.

After all, the sub-contractor does not want to give the impression that he will be 'difficult' and in any event it is he who is in the stronger position when the main contractor is being pushed to complete and he asks the sub-contractor to liven up his installation so that he, the main contractor, can take out his temporary wiring, commission lifts and mechanical services and use the installation for the hundred-and-one-jobs with which he is faced as the project draws to a close.

There are occasions when it may be better *not* to make an issue of a matter at the time one enters into a contract.

Dates for possession and completion

The sub-contractor's period/s for carrying out his work will, to a considerable extent, be dependent upon the conditions of the main contract. Clause 23 of SF 80 reads:

23.1 On the Date of Possession possession of the site shall be given to the Contractor who shall thereupon begin the Works, regularly and diligently proceed with the same and shall complete the same on or before the Completion Date.

23.2 The Architect may issue instructions in regard to the postponement of any work to be executed under the provisions of this Contract.

And the sub-contractor's obligation as stated in clause 11 of NSC/4 reads:

11.1 The Sub-Contractor shall carry out and complete the Sub-Contract Works in accordance with the agreed programme details in the Tender, Schedule 2, item 1C, and reasonably in accordance with the progress of the Main Contract Works but subject to receipt of the notice to commence work on site as detailed in the Tender, Schedule 2, item 1C, and to the operation of clause 11.2.

The wording in clause 11 of DOM/1 is essentially the same.

The date when the *main contractor* takes possession of the site is probably of no great concern to the sub-contractor unless dates which he has been given to work to are linked to the date for possession.

Proceeding regularly and diligently

The sub-contractor will, however, be more interested in the main contractor's date for completion because his own completion period will almost certainly have been related to the main contractor's period. Furthermore, as the sub-contractor is liable to 'observe, perform and comply with all the provisions of the Main Contract' (cl. 5.1.1 in both NSC/4 and DOM/1) he will wish to know the meaning of the words 'regularly and diligently' which appear in the main contract. It has been suggested that the words merely stress what would otherwise be an implied term of the contract, namely, that the main contractor should show that degree of despatch and exertion which would reasonably be expected of his calling (Walker-Smith 1971).

Practical completion of the works

The sub-contractor should understand the difference between 'the Date for Completion' and 'Practical Completion of the Works'. Practical completion occurs 'when in the opinion of the architect Practical Completion of the Works is achieved' and he issues a certificate to that effect (SF 80, cl. 17.1). From both the main contractor's and the sub-contractor's standpoints practical completion is a very important event because it marks the start of the Defects Liability Period. The sub-contractor's liabilities for defects, shrinkages, etc. are set out in clause 17 of SF 80 which reads:

17.1   When in the opinion of the Architect, Practical Completion of the Works is achieved, he shall forthwith issue a certificate to that effect and Practical Completion of the Works shall be deemed for all the purposes of this Contract to have taken place on the day named in such certificate.

17.2   Any defects, shrinkages or other faults which shall appear within the Defects Liability Period and which are due to materials or workmanship not in accordance with this Contract or to frost occurring before Practical Completion of the Works, shall be specified by the Architect in a schedule of defects which he shall deliver to the Contractor as an instruction of the Architect, not later than 14 days after the expiration of the said Defects Liability Period, and within a reasonable time after receipt of such schedule the defects, shrinkages, and other faults therein specified shall be made good by the Contractor and (unless the Architect shall otherwise instruct, in which case the Contract Sum shall be adjusted accordingly) entirely at his own cost.

17.3   Notwithstanding clause 17.2 the Architect may whenever he considers it necessary so to do, issue instructions requiring any defect, shrinkage or other fault which shall appear within the Defects Liability Period and which is due to materials or workmanship not in accordance with this Contract or to frost occurring before Practical Completion of the Works, to be made good, and the Contractor shall within a reasonable time after receipt of such instructions comply with the same and (unless the Architect shall otherwise instruct, in which case the Contract Sum shall be adjusted accordingly) entirely at his own cost. Provided that no such instructions shall be issued after delivery of a schedule of defects or after 14 days from the expiration of the Defects Liability Period.

17.4   When in the opinion of the Architect, any defects, shrinkages or other faults which he may have required to be made good under clauses 17.2 and 17.3 shall have been made good he shall issue a certificate to that effect, and completion of making good defects shall be deemed for all the purposes of this Contract to have taken place on the day named in such certificate (the 'Certificate of Completion of Making Good Defects').

17.5   In no case shall the Contractor be required to make good at his own cost any damage by frost which may appear after Practical Completion, unless the Architect, shall certify that such damage is due to injury which took place before Practical Completion.

The sub-contractor is linked to the contractor in respect of practical completion and liability for defects in clauses 14 of NSC/4 and DOM/1. The clauses of the sub-contracts are similar in their effect. NSC/4 reads:

14   Practical completion of Sub-Contract Works-liability for defects

14.1   If the Sub-Contractor notifies the Contractor in writing of the date when in the opinion of the Sub-Contractor the Sub-Contract Works will have reached practical completion, the Contractor shall immediately pass to the Architect any such notification together with any observations thereon by the Contractor (a copy of which observations must immediately be sent by the Contractor to the Sub-Contractor).

14.2   Practical completion of the Sub-Contract Works shall be deemed to have taken place on the day named in the certificate of practical completion of the Sub-Contract Works issued by the Architect under clause 35.16 of the Main Contract Conditions or as provided in clause 18.1.2 of the Main Contract Conditions.

14.3   Subject to clause 18 of the Main Contract Conditions but without prejudice to the obligation of the Sub-Contractor to accept a similar liability to any liability of the Contractor under the Main Contract to remedy defects in the Sub-Contract Works, the Sub-Contractor shall be liable to make good at his own cost and in accordance with any instruction of the Architect or direction of the Contractor all defects, shrinkages and other faults in the Sub-Contract Works or in any part thereof considered necessary by reason of such defects, shrinkages or other faults due to materials or workmanship not in accordance with the Sub-Contract or due to frost occurring before the date of practical completion of the Sub-Contract Works.

14.4   Where the Contractor is liable under the Main Contract to make good defects, shrinkages or other faults but the Architect instructs that it shall not be entirely at his own cost the Contractor shall grant a corresponding benefit to the Sub-Contractor to the extent that such benefit is relevant and applicable to the Sub-Contract Works.

14.5   The Sub-Contractor upon practical completion of the Sub-Contract Works shall properly clear up and leave the Sub-Contract Works, and all areas made available to him for the purpose of executing those Works and, so far as used by him for that purpose, clean and tidy to the reasonable satisfaction of the Contractor.

These clauses in the main contract and the sub-contracts are important because, whilst it appears onerous, it introduces the Defects Liability Period and so marks the beginning of the end of the contract. Nevertheless, the contractor and the sub-contractor are liable for defects which appear during the defects liability period and they must make them good at their own cost.

Alternative clauses
The sub-contractor should be on his guard regarding alterations to contract conditions which may be proposed by the contractor and regarding the deletion of 'alternatives' which occur in most standard forms of contract.

Alterations to conditions should be kept to a minimum. An apparently trivial alteration to one clause can affect others to a serious extent. On the other hand, where the contract conditions require that one or the other alternatives in a condition should be deleted it is important that there is no doubt about which of the *alternative* sub-clauses apply.

The use of non-standard forms of contract

Sub-contractors are well advised to eschew non-standard forms of contract because they are frequently drafted by lawyers on instructions from, and for the benefit of, the *other* party to the contract. There may, however, be occasions when the sub-contractor is so in need of work in order, say, to keep his work-force employed that he is prepared to consider the use of a non-standard form.

In this event the sub-contractor may find it advisable to have a standard form such as NSC/4 or DOM/i to hand so that he is able to compare the conditions contained in the non-standard form with those in the standard form which has, he knows, been drafted by the JCT and has been agreed by his trade association.

The sub-contractor may be able to obtain some idea of the fairmindedness of the main contractor or client from the amount of information about the contract conditions which is volunteered when inviting tenders for the sub-contract work. Did he, for instance, state:

(a) the form of contract which will be used;
(b) whether the tender should be 'firm' or 'fluctuating';
(c) the way in which fluctuations will be applied;
(d) the defects liability period;
(e) the possible start and completion dates;
(f) the amount of damages;
(g) the percentage of the value of works executed to be retained while works are in progress during the defects liability period;
(h) what, exactly, is required of the sub-contractor by way of insurance provisions;
(i) whether the drawings, bills of quantities and any other relevant documents are available for inspection;
(j) Any phasing of the main contract works which would affect the way in which the sub-contractor has to carry out *his* work?

The sub-contractor may like to use the above questions as a check-list when putting his tender together. It is so easy when visiting a site in the rain or an architect's office where you have obviously come at the wrong time (even though you came by appointment) to rush through what you came to do and miss half the items. A check-list makes it much less likely that you will do this.

Most of the items in the check-list are self-explanatory but comment should be made on one or two points.

Item (a) is fundamental; the sub-contractor may be well advised *not* to enter into a contract if DOM/1 will not be used.

Item (b) is a vital question which one would be unlikely to overlook but nor should one ignore item (c) because if the formula method is *not* to be used one should make provision in the tender for the difference in cost between *actual cost* to the sub-contractor and the sum which will be recovered. Experience shows that the list of basic prices is never complete and that contractors are unlikely to recover the full cost of increases.

The sub-contractor may find the main contractor is not able or prepared to give precise details of start and completion dates (item (e)) if the sub-contractor is to be a domestic sub-contractor.

Walker-Smith (1971) has, in fact, suggested 'that it may not be desirable to commit the parties to firm dates in view of the often necessary changes that occur in the progress of building works' and it may be thought sufficient to agree as a legal obligation the period of the sub-contract works and *preliminary notice required* to start those works.

Before entering into a contract the sub-contractor should be satisfied that sufficient of the drawings and other documents referred to in item (i) are available so that he will be able to carry out and complete the sub-contract works without disruption to the progress of the works. He should be particularly cautious in respect of the implications of item (j).

A key decision

Entering into a contract is the sub-contractor's *raison d'être* but it also represents a crucial decision which should be made only after all the relevant facts have been considered. The climate for negotiation is usually more equable before than after the final commitment so he should take care to obtain the best possible terms whilst he has the opportunity.

References

Walker-Smith, D. (1971) *The Standard Forms of Building Contract*. Knight, Tonbridge.

Further reading

Bowden, G. F. & Morris, A. S. (1978) *An introduction to the law of Contract and Tort*. The Estates Gazette, London.
Turner, D. F. (1971) *Building Contracts, A Practical Guide*. Codwin, London.
Manson, K. (1981) *Building Law for Students*. Cassell, London.

Chapter 7

Production planning, budgets and control

Production planning commences with the preparation of the specialist sub-contractor's estimate. Without a basic plan the contractor is unable to ascertain the time he will require to carry out the works so that he can arrange his finances and he will be unable to make a realistic estimate of the cost of supervision, site management and plant use.

It was, therefore, appropriate to introduce production planning in Chapter 4 and the comments made in that chapter will not be repeated here. A bar-chart programme of the type used by main contractors for pre-tender planning was also discussed in Chapter 4.

Such plans are readily understood by the people who have to use them so why bother with network and critical-path planning?

The answer is that bar-charts are good for *presentation* of planning data but they do not lend themselves so readily to the *preparation* of programmes especially when the project is complicated or there are numerous operations which have to be slotted together.

Use of networks

Networks grew out of the space-race when the Americans discovered that they needed a method of planning and controlling the thousands of operations or activities which had to proceed both sequentially and concurrently if they were to succeed with their programme. They needed a method which showed the dependency of one operation upon the other. Critical path planning was one of a number of programme evaluation and review techniques (PERT for short) which was developed.

The first step in the plan is to break the project down into its component operations. In the table in Fig. 4.4 fourteen electrical installation operations were identified; fifteen if one includes site supervision and management. In practice it may be necessary to break down some of those 'operations' still further because *an essential feature of an operation is that it must run without interference from another operation.*

The object of network analysis is to arrange the operations in a logical sequence or several concurrent sequences so that the dependence of each operation upon the last can best be seen.

In order to do this each operation is represented in the programme by an arrow above which it is usual to write the operation title:

Fit convectors
———————➤

If, as might be the case on the Excell Offices project, the convectors were situated on several floors and the fixing of each floor's convectors was an independent operation the total time required for all the convectors would be broken down accordingly and the operations would read:

Fit convectors
(ground floor)
—————————➤
(6) duration

Beneath the arrow is written the duration. The duration can be in any units but it is essential that the same unit is used throughout the programme, i.e. if the planner starts with days as the unit all the durations must be expressed in days.

There is no significance to the length of the arrows in the normal network. The arrows are joined by nodes; the tail of the arrow indicates the beginning of the operation and the head the finish.

The nodes can be divided into three sections, e.g.

or into quadrants (hot-cross buns) e.g.

Hot-cross buns are probably used less since up-dating (of which more anon) has been carried out by computer but they are more convenient for networks which are up-dated manually and will be used for these examples. The nodes contain information about the timing of the operations which precede and follow the nodes:

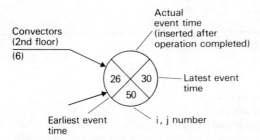

The *earliest event time* is the earliest time by which all the operations heading towards the node will have been completed. The time which will be inserted in the left-hand quadrant will be that of the arrow with the greatest number at its head.

The *latest event time* is the latest time by which all the operations heading towards the node must have been completed. The time which will be inserted in the right-hand quadrant will be that of the arrow with the greatest number at its tail when working backwards through the programme subtracting operation durations from the latest event times of the nodes to the right of the node whose latest event time is being calculated.

There are some other terms which will have to be explained later but with the information given above it is possible to construct a simple network. The logic of networks is that an operation cannot start until the preceding operation/s has/have been completed. Specialist sub-contractors are well aware that their work in connection with building projects is not carried out in isolation so it will be appreciated, for instance, that the mains switchboards and meters, operation 1 in Fig. 4.4, cannot start until the main contractor has built the wall on which the switchboards and meters will be fixed.

Reference to Fig. 4.3 shows that the internal partitions, services, wall and ceiling finishes follow closely after each other. Assuming that for technical reasons the main contractor has planned to build the internal partitions from the bottom to the top of the building the sequence of operations might be as shown in the schedule shown in Table 7.1.

As for the mechanics of preparing networks, some elementary rules are:
(a)   use a large piece of paper;
(b)   work in pencil;
(c)   always have an eraser available;
(d)   never be afraid of altering the network as operations which have been forgotten are remembered or there is an obvious fallacy in one's 'logic'.

It is now possible to express the operations listed in the schedule as a network, insert event times, critical paths, etc. and discuss 'float' and similar matters.

Using the symbols and principles discussed above the schedule of operations

Table 7.1 Schedule of operations for ground floor

	Duration* (days)
Construct internal partitions	15
Electrical mains and cables	8
Engineering services carcassing	10
Plumbing carcassing	7
Wall and ceiling finishes	23
Lighting and power outlets	4
Engineering services second fixing	8
Plumbing fittings	6
Joinery ground and bearers	6
Ceiling hangers	4

The durations for the operations would vary with each floor

contained in Table 7.1 may be translated into the network for part of the programme shown in Fig. 7.2. How is the network constructed?

The object of network analysis is to arrange the operations in a logical sequence and with the main requirement that an operation must run without interference from other operations.

Working from left to right, the network starts with the building of the internal partitions on the ground floor. A requirement of the programme is that these partitions must be complete before the services, carpentry works and ceiling hangers can commence at the node numbered 38; the number in the bottom quadrant of the node.

The matter of notation is discussed below.

All the operations immediately to the right of node 38 run concurrently without interfering with each other. All these operations, except the internal partitions on the 1st floor, must be complete before the wall finishes and subsequently the ceiling finishes on the ground floor can commence. Only when these finishes, which make up operations 55–57 and 57–59, are complete can the 'second fix' services operations commence. These operations must, in their turn, be complete before the operation which starts at node 75 can commence. We are concerned with only part of the programme so our network stops at that operation.

If we return to node 38 we see that the internal partitions on the first floor must be complete before the services and other carcassing work on that floor can commence and the whole cycle which we have just followed through on the ground floor can run its course.

The event times for nodes 59–72 in Fig. 7.1 are shown below that figure.

Event times and dummies

Reference was made above to event times.

The durations in the network are in working days. The earliest event time at node 35 is therefore day 190. Operation 35–38, the internal partitions on the ground floor, has a duration of 15 days so that the earliest event time for subsequent operations is day 205 (190 + 15). All these operations must be complete before the wall finishes can commence but as they are all independent operations it is essential that they have independent identities – they cannot all be called operation 38–57.

To facilitate identification nodes have been introduced between nodes 38 and 55 (39, 43, 47, 49 and 53) and *dummy operations* have been created between the intermediate nodes and node 55.

Dummy operations are drawn with broken lines. They are essential for the construction of the network as they indicate the dependency of the operations upon each other but they have no duration.

The earliest event time at node 55 is day 215 because this is the earliest time by which all the operations headed towards node 55 will be complete – the completion day of operation 38–39, the operation with the longest duration.

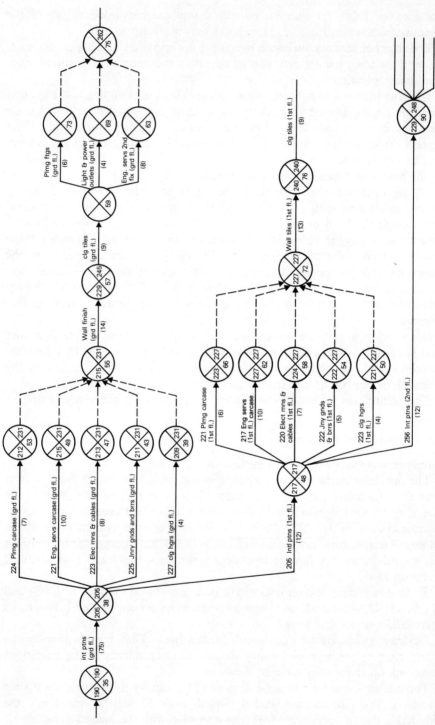

Fig. 7.1 Network for part of programme

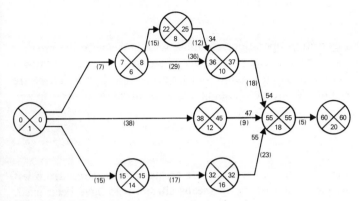

Fig. 7.2 Network illustrating converging paths

Converging paths

Although it does not occur in the network in Fig. 7.1 it is quite common for several paths of operations (as distinct from dummy operations) to converge on a node. In this event it is advisable to pencil-in the earliest event time of each operation near its head so that the path with the longest duration can easily be selected and its value inserted in the earliest event time quadrant (see Fig. 7.2).

If we revert to the Excell Offices network it can be seen that most of the earliest event times have been inserted in the nodes. Some have been left out in order to provide an exercise in event time calculating. The missing times are given at the end of the chapter. Having inserted all the *earliest* event times it is necessary to calculate the *latest* event times for insertion in the right-hand quadrant of each node.

In the network in Fig. 7.2 the earliest and latest event time for the terminal node must obviously be the same. We can, therefore, start with a latest event time at node 20 of day 60. From this is deducted the duration of operation 18–20 and subsequently the duration of each of the operations 10–18, 12–18 and 14–18. The same procedure is adopted at nodes 10, 12 and 14, and so on back through the network. Where two or more operations converge at a node, as at node 6, the latest event time of each operation can be pencilled-in at the tails as shown.

The same procedure can be seen in more detail on the Excell Offices network in Fig. 7.1.

Having gone to the trouble of calculating these event times, what can we learn from the network?

Critical path

The most important piece of information is the critical path. This runs through the operations joined by nodes with earliest and latest event times which are the same. It is usual to indicate the critical path by making it thicker or making two short strokes through the line near the head of the arrow.

Float

It follows that the *least* critical operations are those with nodes with the greatest difference between earliest and latest event times. This spare capacity is known as 'float'; the operation can *float* on without affecting the critical path. There are several types of float but for simple planning purposes these need not concern us too much.

Notation

It can be seen that the numbers in the nodes are greater as one moves from left to right along the network but that by no means all numbers have been used. Numbers are not inserted in the nodes until the network is to all intents and purposes complete. Some numbers are left out to make it possible to revise the network by, for instance, inserting an operation without needing to renumber.

Interpreting networks

It is possible to obtain more information from a network than merely the critical path.

If one continues to view from the standpoint of the electrical contractor it may be seen that at the carcassing stage of the services on the ground floor of the Excell Offices project, (see Fig. 7.1) the electrical mains and cables require 8 days to complete compared with 10 days required for the services carcassing. It can be seen, therefore, that unless the services carcassing is accelerated or the electrical works are delayed, the electrical works will not become more critical than the services carcassing works. Furthermore, as the difference between the earliest and latest event times at node 55 is 16 days, day 231 minus day 215, we can see that even if the services carcassing were to be delayed by up to sixteen days the whole of the network between nodes 38 and 57 would not become critical.

Obviously this is useful information for the sub-contractor. No one would suggest that a sub-contractor should not carry out his work as expeditiously as possible; indeed, he has a contractual obligation so to do, but he cannot be accused of causing delay unless delay has actually occurred and this is not so in this instance. Clearly an ability to read and understand a network enables a sub-contractor to discuss programmes with the main contractor on more equal terms.

Examination of the network shows there is much less float for the electrical sub-contractor on the first floor. At present the critical path runs through the engineering services carcassing work, wall finishes and ceiling tiles (ops 48–62, 62–72 and 72–76). If the electrical sub-contractor were to take more than 10 days the course of the critical path would be on the electrical mains and cables and the electrical sub-contractor would be causing delay.

Sub-contractors should review their programmes at least monthly and often more frequently. If we assume that the Excell Offices network shown in Fig. 7.1

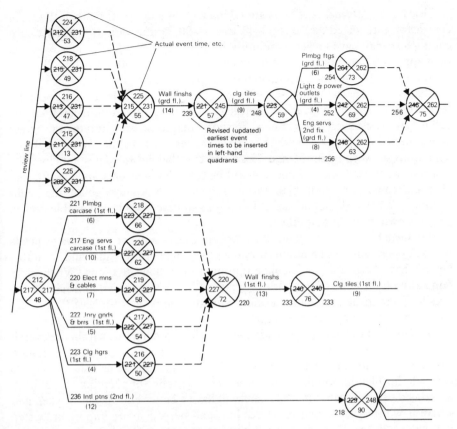

Fig. 7.3 Updated network for part of programme

is being reviewed at about day 225 we might find the actual progress to be as shown in Fig. 7.3. Progress appears to be approximately ten days behind schedule on some of the works and ahead on others.

By day 220 it will be possible to say quite accurately when the plumbing carcase and ceiling hangers on the ground floor will be completed so that the date at node 57 can be given with reasonable confidence. The revised earliest event times at node 59 and subsequent nodes on that path will be altered as shown in Fig. 7.3, the revised dates being inserted to replace the original dates. A copy of the original plan must be kept as a record because it may become important evidence later.

At node 72 progress appears to have been better. The main contractor and sub-contractors may have benefited from their experience on the ground floor. It appears that the operatives fixing the ceiling hangers, who perhaps were late arriving on site, were put to fixing the hangers on the first floor before those on the ground floor. No doubt this was because the critical path runs through the operations between nodes 48 and 76.

Clearly, the internal partitions are taking less time to construct than the main contractor anticipated (see ops 38–48 and 48–90) so the various specialist sub-contractors may have greater continuity in their carcassing work, floor by floor, than appeared to be likely in the original plan. Operation 90–? should follow up operation 38–47 without more than a day or two's delay.

Changes in plan such as those suggested above are inevitable and can generally be accommodated by both main contractor and sub-contractors. A study of the earliest and latest even times at node 76 shows that if the present rate of progress is maintained the project could be completed ahead of schedule even though the operations between nodes 55 and 75 are ten days behind the dates anticipated.

From what has been said above it will be clear that networks are a useful tool of management for both planning and control. A knowledge of the main contractor's network can be useful to the sub-contractor as it will enable him to plan his own work to best effect.

It will enable him to work out the number of man-hours or man-days he needs for each operation that he has to carry out at the time of his tender and from this he will have been able to calculate optimum operation durations. When he is familiar with the main contractor's programme he is able to ascertain where pressures lie – the main contractor's critical path and the paths with least float – and concentrate his resources accordingly.

He will also be able to prepare his own, detailed, programme within the overall framework of the main contractor's programme. If he does this in consultation with his site team and perhaps even obtain the main contractor's site manager's assent to his plan he will have done well. The discipline of planning will in itself have been useful because management will have a better knowledge of when and where it will require which resources but, and perhaps more important, a step will have been made towards the 'teamwork' to which we all so often pay lip-service but so seldom take seriously.

Manpower planning

The sub-contractor may use the networks discussed above to assist him with planning his labour force if he writes the manpower requirement under the arrow representing the operation:

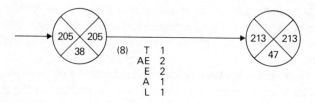

The initials refer to the grades of operative listed in the schedule of labour costs issued by the electrical contracting organizations. The number '(8)' refers to the duration of the operation.

Use of bar-charts

On a comparatively simple project such as the Excell Offices it should not be too difficult to relate the operations to the overall time-scale of the project but on more complex projects the sub-contractor may find it easier to prepare an elementary bar-chart of his and the main contractor's operations and produce histograms of the time to be spent by his operatives.

Figure 7.4 shows a small part of bar-chart with, below, the histogram which relates to it in order to illustrate the method which is used. The part of the programme where the electrical sub-contractor is employed on three operations is shown in Fig. 7.4 and it is assumed that his estimate of his manpower requirements, taken from his tender for the works, provides for:

(a) *switchroom*: two electricians and one apprentice;
(b) *riser*: one electrician, one apprentice and one labourer;
(c) *ground floor mains and cables*: three electricians and one labourer.

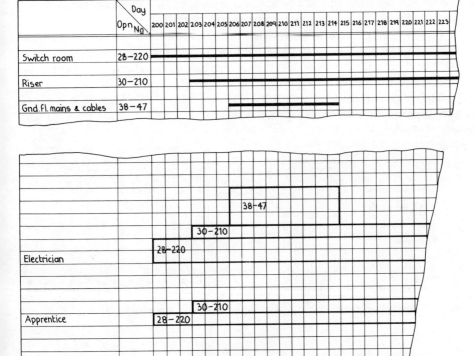

Fig. 7.4 Part of bar-chart with histogram below showing manpower requirements

From the histogram section of Fig. 7.4 it can be seen that what one does is to regard each square on the table as an 'operative/day' block and so build up heaps of blocks representing the number of each type of operative on the site on any day. The operation numbers of each block are written on the appropriate block.

It can be seen in the example that between day 206 and day 214 a peak of six electricians will be required on the site. Needless to say, the aim of manpower planning is to keep one's labour force as constant as possible. With only three operations running one has limited flexibility but if one had more running one would attempt to 'level' out the peaks into the valleys.

This can usually be done by borrowing operatives from operations with 'float' (non-critical operations), to boost the operations on the critical path.

It can be seen in Fig. 7.4 that each grade of operative must be given its own histogram. In practice it is usually advisable to use a separate piece of ruled paper for each grade and join the sheets together, one above the other, when one knows what the profile of each grade will be. The above exercise may appear unnecessarily complicated for the normal run of jobs but for larger jobs the use of histograms can be useful as a means of determining the labour strength one will need, reducing the peaks and troughs as far as possible.

For less complicated jobs involving just an operative or two or, perhaps, an operative and mate, one can write the number of operatives in the time-bar of the bar-chart and add these together at the foot of the chart. This simple but often perfectly adequate method can be seen in Fig. 7.5.

The price one pays for simplicity is loss of adaptability because it is not so easy to 'resource level' using the latter method.

	0	1	2	3	4	5	6	7	8	9	10	11	12	13	14	15	16	17	18	19
Circuit conduits	4	4	4	4	4	2	2	2	2	2	2	2	2	2	2					
Sub-main cables						2	2	2	2	2	2									
Distribution gear												2	2							
Circuit cables														2	2	2	2			
Accessories																		2	2	
Luminaires																			4	4
Men on site	4	4	4	4	4	4	4	4	4	4	4	4	4	4	4	4	4	4	4	

Fig. 7.5 Method of showing manpower requirements for simple jobs

Short-term plans

The methods described above are of primary importance to the main contractor but of rather less importance to specialist contractors because many of their contracts do not involve very many operations.

The methods are, however, applicable for the specialist contractor's work and informal programmes are often very helpful in planning work for the forthcoming month, week or even day.

Pro forma bar-charts may be prepared which can have the appropriate time intervals inserted.

These bar-charts may be used as a means of instructing supervisors regarding the specialist contractor's plans in the context of the main contractor's programme. There is a tendency to believe that bar-charts should have a polished presentation. This is not so; they are primarily management tools and it is their *function rather than their presentation which matters.* Indeed, there is sometimes reason to suspect the well-presented programme – it may have been prepared for public relations purposes which have little to do with the actual progress of the works.

Planning and control for short duration jobs

Figure 7.6 shows a pro forma for a job assignment sheet which is suitable for short duration operations or 'jobs'. The aim is to provide all the information for planning, controlling and costing the individual jobs.

The pro forma is fundamentally a working 'tool' which tells the operative what has to be done, where, when and how it is to be done and what he is to do when the job has been completed.

Responsibility for planning and method rests with the manager and if he is to complete the pro forma properly he must have thought through what the operatives have to do before he gives his instructions. There is provision for target time and 'actual' time if an incentive payment scheme is in operation. The 'total time' may be used as the basis for charging the cost of the works to the client if this is necessary.

In addition to use by specialist contractors on major projects the job assignment sheet may be used for, say, maintenance work in a hospital or by specialist contractors who have jobs to carry out in different places in a building. The plans and sections of the building shown on the job assignment sheet are important for locating the job site. The plans and sections may be prepared for a particular project and photocopied onto a batch of pro forma for use on that project.

As this chapter has so far been concerned with *production* planning and control at 'contract' or 'project' level, it is logical to use the contract as a starting point to discuss *budgetary* planning and control but it will be apparent later in the chapter that budgetary planning must start 'at the top' of the organization.

Monitoring the programme
Planning for building work is in itself important because planning is often described as 'building on paper' and has the advantage that if one makes a mistake one can remedy it more cheaply than is possible with actual building work. Planning is, however, only half the story – the other half is *control*.

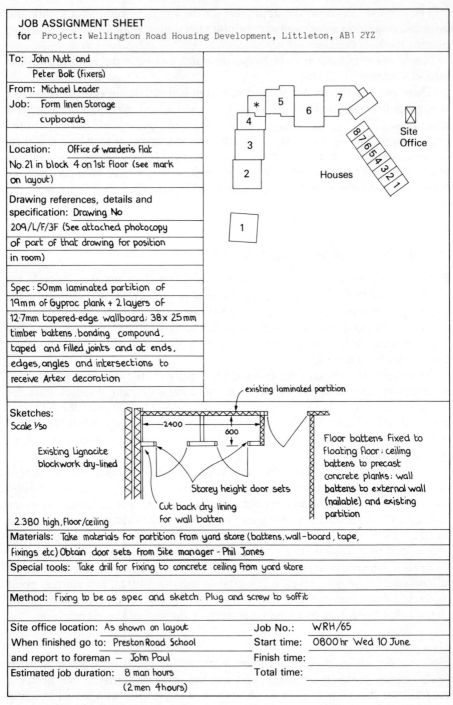

JOB ASSIGNMENT SHEET
for Project: Wellington Road Housing Development, Littleton, AB1 2YZ

To: John Nutt and
Peter Bolt (fixers)

From: Michael Leader

Job: Form linen storage
cupboards

Location: Office of warden's flat
No. 21 in block 4 on 1st floor (see mark
on layout)

Drawing references, details and
specification: Drawing No
209/L/F/3F (See attached photocopy
of part of that drawing for position
in room)

Spec: 50mm laminated partition of
19mm of Gyproc plank + 2 layers of
12.7mm tapered-edge wallboard; 38 x 25mm
timber battens, bonding compound,
taped and filled joints and at ends,
edges, angles and intersections to
receive Artex decoration

Houses

Site Office

Sketches:
Scale 1/50

existing laminated partition

Existing Lignacite
blockwork dry-lined

2400
600

Storey height door sets

Cut back dry lining
for wall batten

2.380 high, floor/ceiling

Floor battens fixed to
floating floor; ceiling
battens to precast
concrete planks; wall
battens to external wall
(nailable) and existing
partition

Materials: Take materials for partition from yard store (battens, wall-board, tape,
fixings etc) Obtain door sets from Site manager - Phil Jones

Special tools: Take drill for fixing to concrete ceiling from yard store

Method: Fixing to be as spec and sketch. Plug and screw to soffit

Site office location: As shown on layout	Job No.: WRH/65
When finished go to: Preston Road School	Start time: 0800 hr Wed 10 June
and report to foreman — John Paul	Finish time:
Estimated job duration: 8 man hours	Total time:
(2 men 4 hours)	

Fig. 7.6 Report pro forma for short-duration jobs

Definition of control

Control may be defined as regular and frequent comparison of actual performance with anticipated performance so that undesirable trends away from the standard (the norm) can be detected and corrected at an early stage. It follows that control must go hand-in-hand with planning because unless one has a plan one has no standard by which to measure one's performance. Specifications for controls may be identified.

The controls must be economical. If the cost of running the production control department is greater than the savings which can be effected by not controlling production, management should consider alternative methods of achieving its aim.

Controls must be meaningful and appropriate. The control technique which is suitable for quality control may be, indeed almost certainly would be, inappropriate for cost control. Controls should not be concerned with too-fine details. The question the manager should ask himself is: 'What is the minimum information we need to know in order to control?'

Controls must be congruent. The manager should remember that controls are concerned with detecting and correcting trends and in order to do this he needs control data which indicates trends. Trends cannot, by definition, be measured to the *n*th degree.

Controls must be timely. Timing is more important than a high level of accuracy because if controls are not carried out regularly and frequently they will be ineffective.

Controls should be simple. Controls should be a practical rather than a 'window dressing' exercise. Building is a labour-intensive industry and as controls must be acceptable and understood by the personnel who are closely concerned, the control 'standard' must be simple.

Controls must be operational. The form of control must be related to the operation which is to be controlled and the control method to be used must, itself, be capable of being operated. The specification that controls should be simple, above, applies to this specification and, indeed, all the specifications are inter-related. Production and cost-control techniques must be integrated within the builder's management system.

The standards referred to in the definition given above are based on normal operating efficiency. For production and cost-control purposes the standards units may be expressed in terms of time and cost. It follows from this that the 'predetermined standards' will be time programmes and budgets.

Without a programme or a budget, production and cost control are impossible because there is nothing to measure actual progress, expenditure etc., against in order to detect and correct undesirable trends.

In two words we are concerned principally with controlling time and money.

Budgetary planning and control

Production planning and control are concerned primarily with time and progress but from what has been said above it is evident that progress may be measured in various ways.

Cash-flow and profitability are matters of concern to all contractors so the use of cash as a 'yardstick' clearly has much to recommend it. Furthermore, cash is an element which is common to all aspects of the contractor's activities.

Budgetary planning and control make it possible for the contractor to have a common yardstick for all his projects and supportive departments.

The principal benefits of a formalized budgetary control system can be summarized as:
(a) setting out in financial terms the objectives of the organization;
(b) providing yardsticks by which to measure efficiency at all levels in the organization;
(c) demonstrating the extent by which results vary from anticipation (the variances);
(d) providing a guide for corrective action;
(e) making possible control at the 'top' (level of management) with delegated responsibility.

What is a budget?

One answer to that question is a pre-determined plan, expressed in financial terms, covering all phases of a business. A budget should state specifically what is expected of the company in terms of profitability and how the plan is to be achieved. It should be a well-conceived plan which ensures the desired level of profitability allied to all the resources available to a business.

It will immediately be apparent to the contractor that this description fits quite closely the term 'planning' in the form that he plans (programmes) his individual projects although he may not, perhaps, attempt to put costs to his plan in quite the way envisaged above.

Budgetary control is not an alien concept to the specialist sub-contractor but rather an extension of his contract planning to embrace not just individual contracts but his whole business.

Standards and variances

Budgets provide the *standard costs* against which *actual costs* may be measured. Standards are based on normal *operating efficiency*.

The differences between the standard costs and the actual costs produce *variances* which may be expressed in terms of *price* and usage to make up the *total variance*.

Example
Assume
Unit cost of item (heating element, m² of blockwork, etc.) has been calculated as:

	(£)
Labour	6.58
Materials	4.60
On-cost	2.24
	13.42

Assume number of units = 200
Therefore *standard cost* = 200 × £13.42 = £2,684
Assume unit cost varied in price due to increase in labour cost (or material cost) of £0.78 each to £13.42 + £0.78 = £14.20; the *price variance*
Assume the quantity of units is varied because 221 units are actually used; the *usage variance*.
Total (adverse) variance would be;

Actual cost; 221 × £14.20	= 3,138.20 minus
Standard cost; 200 × £13.42	= 2,684.00
Total variance:	= £454.20

Budgetary planning and control for individual contracts

The aim of budgetary planning and control is to allocate the expenditure on the various operations to the time when the expenditure will occur. If an integrated management system is adopted, budgetary planning and control should be part of the whole system. Furthermore, as many contractors incorporate a productivity-based Incentive Payment Scheme (IPS) within their management system, any project-based budgetary planning system should be designed with this additional function in mind. Incentive schemes are discussed in Chapter 10. A comprehensive budgetary system should be concerned with *all* aspects of costs; labour, plant and materials, but as many contractors do not attempt to use budgetary planning methods for controlling plant and materials costs, discussion will, initially, be restricted to controlling the most variable of a contractor's costs, namely labour.

To plan and control the cost of labour it is necessary to be concerned with 'estimate' and 'expenditure'. The labour content of the contractor's estimate should be used for budgetary purposes.

Assuming the labour estimate for an operation is £480 and the programme indicates that it will take twelve days to complete the operation, the daily budget is

$$\frac{£480}{12 \text{ days}} = £40/\text{day}.$$

A budget for a day is of limited use for budgetary planning and control purposes but it is a useful unit as it can easily be converted to a monthly or weekly budget by multiplying by the number of days in the month or week as the case may be. The average month is usually about twenty-two days but they vary considerably month by month due to bank holidays, Christmas period, etc.

When considering budgets which will be of use for both budgetary control and IPS purposes a decision must be made regarding the most appropriate period of time for the two purposes.

Experience on construction projects has led the majority of contractors to discover that incentive payments schemes require payments to be made weekly. if they are to work properly, so the 'week' is usually the unit of time used for both cost control and IPS purposes.

When the frequency has been decided the labour value calculated above can be allocated to a time-scale. Figure 7.7 adapts the chart shown in Fig. 7.4 to incorporate budget values. It is now possible to see what the weekly labour cost should be for each operation and by reading the totals at the foot of the programme, for the project as a whole, both weekly and cumulative totals being given. The budgets are only targets and before one can ascertain if the operations are being carried out profitably it is first necessary to *value* the work which has been completed and to calculate the *actual* value of the work which has been completed to date.

Calculating comparable values
Valuing the work actually carried out usually requires that some sort of measurement is made at the end of the period in question. This should not be too difficult to do particularly if it can be related to, say, a repetitive item of work.

A review of progress carried out at the end of week ending 26 April, using Fig. 7.7, might indicate that 40 per cent of the work to the riser had been completed. If the total labour value of the completed riser is £3,125 the value of work completed to date would be 40 per cent of £3,125 = £1,250. From this sum must be deducted the value of work previously completed, £890 in this instance, to give a labour value for week ending 26 April of £360.

Comparing budgets and values
Where does this lead? What is the point of preparing budgets and measuring the value of work each week? One important reason is that comparison of budgets with actual values enables the sub-contractor to know if the work is on programme.

Experience shows that productivity and progress can most readily be measured if they are converted to £.p. terms. Comparison of the budgets with the actual

W/E	5/Apr					12/Apr					19/Apr					26/Apr					3/May					10/May					
Day / Opn no	200	201	202	203	204	205	206	207	208	209	210	211	212	213	214	215	216	217	218	219	220	221	222	223	224	225	226	227	228	229
Switch room 28–220				350		325		350		370		350		380		350		360		350						350				
Riser 30–210				240		220		320		325		320		345		320		360		320						320				
Gnd Fl. mains cables 38–47						380		330		445		450		—		1		75												

	Budget	Value	Budget	Value	Budget	Value	Budget	Value	Budget	Value	Budget	Value
Weekly total	590	545	1,050	1,025	1,115	1,175	670	795	670		670	
Accumulative total	1,090	1,520	2,240	2,545	3,385	3,720	4,055	4,515	4,725		5,395	

Fig. 7.7 Bar chart with budgets and values

119

values for the switchroom shows that work in the switchroom is ahead of programme whereas work on the ground floor mains and cables was behind programme during both week endings 12 and 19 April, and indeed, overran by 3 days.

In practice, budgets are most useful for long-duration operations because without budgets it is difficult to know, often until too late, whether or not work is on programme.

The budgets in themselves can be used to maintain positive cash-flow because they inform the sub-contractor, in advance, of the amount of money he will need to provide for labour and materials each month on each contract. The values, as measured, can provide the basis of the sub-contractor's interim applications for payment on accounts.

A more sophisticated approach to budgetary planning and control is demonstrated in Fig. 7.8. This Monthly contract report is suitable for specialist sub-contractors undertaking larger projects on which the sub-contractor has nominated or domestic sub-sub-contractors. SF 80 makes provision for such an arrangement which is by no means unusual on major projects.

The monthly contract report is a summary of the monthly budgets.and costs for each of the cost centres – labour, plant, materials, etc. shown in section 1 of the report.

The summary represents the total of all the budgets and costs for the various operations carried out during the month.

The report is designed so that its columns correspond with those on the programme and so that it can be read straight from the programme which is in bar-chart form with budgets for the operations.

The budgets are based on prices prevailing at the time of tender, while the costs are those prevailing when the work is carried out. The budgets and costs are, then, an unrealistic basis for comparison but they do provide a base from which to work. Allowance for the inflation factor can be made separately as shown in section 2 of Fig. 7.8.

If the NEDO price adjustment formula is operating the indices used when making fluctuations and adjustments may be applied to obtain the adjusted budget figures in section 2.

The result will not be precise, but we must not forget that for control purposes we are concerned with trends, rather than precision.

Nor need we concern ourselves with the different increases or decreases between work categories. The lump sum for the month's valuation is all that is needed for control purposes.

Generally, it is not necessary to make the adjustments described above to the individual operations but the method is applicable if it is desired to ascertain the adjusted cost of the operations for, for instance, productivity-based incentive payment schemes.

Sections 3 and 4 provide the descriptive part of the report. They are really check-lists to ensure that action is taken to put in train procedures for the contractor to obtain an extension of time or for him to apply for reimbursement

MONTHLY CONTRACT REPORT

for (contract) Wellington Road Housing Developement

for Period No. 3 ending 31 July 1982

Code ref. WRH/81/8

1 — Budgets and costs summary

Month/pd	June 82 Budget	June 82 Cost	(3) July Budget	(3) July Cost	(4) Aug. Budget	(4) Aug. Cost	(5) Sept. Budget	(6) Oct. Budget	(7) Nov Budget	(8) Dec Budget
Labour	2,200	2,460	3,600	3,750	4,620		4,756	4,805	4,950	4,900
Plant	1,450	1,590	1,450	1,590	1,750		1,850	1,950	1,850	1,900
Materials	5,900	6,450	6,700	7,206	9,700		9,556	9,320	9,000	9,200
Prelims	6,950	7,100	7,300	7,430	7,300		7,300	7,300	7,300	7,300
Own s-cs	3,650	3,720	5,760	5,850	7,830		7,900	7,800	8,100	8,200
Nom s-cs	4,440	4,520	6,230	6,404	8,820		9,400	9,300	9,300	9,500
Totals	24,590	25,840	31,040	32,185	40,020		40,762	40,475	40,700	41,000

2 — Value and adjusted costs summary

Month/pd	June 82 Adj.bgt.	June 82 Value	(3) July Adj.bgt.	(3) July Value	(4) Aug. Adj.bgt.	(4) Aug. Value
Labour	2,506	2,735	4,050	4,122		
Plant	1,500	1,590	1,620	1,590		
Materials	6,400	6,722	7,225	7,560		
Prelims	7,150	7,360	7,527	7,720		
Own s-cs	3,800	3,908	5,950	6,135		
Nom s-cs	4,550	4,684	6,450	6,674		
Totals	25,906	27,019	32,822	33,801		

3 — Contract progress

3.1 Contract is on /not on* programme

3.2 Contract is 3 days/weeks* ahead of/behind* programme

3.3 Extension of contract period is necessary/not necessary* for 3 days/weeks*

   „   „   „   3.4 was applied for on (date) 10 June for 3 days/weeks*

   „   „   „   3.5 was given on (date)   for / days/weeks*

3.6 Accelerated progress: instructions have/have not* been given to accelerate progress on (date of instruction) /by /days/weeks*

4 — Factors affecting progress

4.1 Previously advised factors: exceptionally adverse weather conditions (29 May)

4.2 New factors: drawings for roof trusses still outstanding, applied for 12 April, fixing planned to commence week 25-13 Sept

*delete as appropriate.

Fig. 7.8 Monthly contract report

because the regular progress of the works has been affected, as the case be. Section 3 facilitates prompt detection of undersirable trends from the standard and makes it possible for action to be taken.

The procedures which must be taken if the contractor is to obtain an extension of time or for him to apply for reimbursement because the regular progress of the works has been affected in accordance with clauses 25 and 26 of the 1980 edition of the Standard Form of Building Contract require considerable attention to detail. These procedures are discussed in Chapter 13. It is sufficient, for the time being, to note that section 3 ensures that any departure from the norm will not pass unnoticed.

Similarly, the requirement for the manager to enter in section 3.5, whether or not an extension of time has been given, is a strong inducement to management to take the appropriate action.

While the contractor has an obligation in clause 25 to 'use constantly his best endeavours to prevent delay in the progress of the works' there is no requirement for him to accelerate progress unless he has been instructed to do so. Section 3.6 provides a check-list for the contracts manager which leaves neither site manager nor contracts manager in doubt with regard to the action which is to be (or has been) taken.

Section 4 in the monthly contract report is intended to bring, and keep, the contractor up-to-date regarding factors affecting progress. The report should be succinct because correspondence, minutes of site meetings and other records will take care of the detail.

The monthly contract record, together with the accompanying programme, makes it possible for the contracts manager to monitor and control cost and progress – his two most important standards.

In the event of undesirable trends being detected the contracts manager has access to the project programme so that he may ascertain which operations are causing problems. This is an example of 'management by exception', an approach to control which identifies exceptions from the norm so that the reason for their exceptional performance may be ascertained and action may be taken.

Budgetary planning and control for all contracts

This is the concern of the management group, managing director or chief contracts manager/engineer. The person or persons responsible will depend on the organization structure of the firm in question. For purposes of this chapter it will be assumed that a management group will control the contracts.

Cost centres

These are determined for each contract using the data contained in the estimate for each contract as the basis. The estimate provides the budgets for a contract and a typical contracts data analysis is shown in Fig. 7.9 which summarizes the data for all the contracts which the specialist contractor has in train.

Columns 1 and 2 identify the contracts within the contractor's management

Sheet 1

Contracts Data Analysis

1 Contract	2 Code ref.	3 Form SFBC etc.	4 Firm Yes/No	5 Start date	6 Period (mths)	7 Cont'ct sum	8 O/hds & pft.	9 Labour	10 Plant	11 Matls	12 Prelims	13 Own SCs	14 NSC & S	15 Remarks
Wellington Road Housing (Contract A)	WRH 80/8	SFBC LA	NO	April'81	17	825,500	55,600 7.22%	66,000	25,600	118,400	123,800	210,600	225,500	
Allbourne Social Centre (Contract B)	ASS 80/9	SFBC LA	YES	Feb'82	10	75,000	6,100 3.85%	5,700	2,300	7,200	10,400	7,600	35,700	
Pendel Road Flats (Contract C)	PRF 80/13	SFBC Private	NO	JUNE'81	14	920,600	48,300 5.55%	127,300	36,700	180,400	139,600	167,200	221,100	
Rodley Hospital Extension (Contract D)	RHX 80/4	SFBC LA	YES	FEB'82	9	150,000	8,700 6.16%	27,900	7,200	21,000	27,600	21,500	36,100	
Pilborough Hospital (Contract E)	PH 81/6	SFBC LA	NO	SEP'82	22	1,472,400	68,900 4.91%	131,700	74,500	155,600	212,500	98,400	730,800	
Folten Comprehensive Sch (Contract F)	FCS 81/11	SFBC LA	NO	FEB'82	18	1,200,700	59,400 5.20%	109,000	71,700	211,400	145,800	142,000	361,400	
Ashley Place Hotel (Contract G)	APH 81/12	SFBC Private	NO	April'82	18	982,000	61,700 6.70%	87,200	39,900	131,700	129,800	221,100	310,600	

Fig. 7.9 Contracts data analysis

system. Column 3 provides for the form of contract which will be used. In the example the contracts employ either the 'local authority' or 'private' editions of the Standard Form of Building Contract.

In effect, column 3 lays down the rules for conduct of the contract. Column 4 requires a direct 'yes' or 'no' answer to the question concerned with whether the contract is firm price or variation of price.

Columns 5 and 6 enable the contracts management group to allocate time to the contract. If the contract period is extended during the progress of the works the extended period would be recorded in column 6, provided the extension had been agreed with the client. Column 7 contains the contract sum.

In this column, as in column 6, the effect of variations and extensions of time would be recorded so that the sum represents the true standard by which to measure performance. Only such cost variations as have been agreed with the client or the main contractor should be entered in this column. Claims which might or might not succeed should not be included.

Columns 8–14 are a break-down of the tender sum. The overhead and profit content of the contract (in col. 8) is most easily extracted from the contract sum if the unit rates (where the contract is based on the traditional bills of quantities) are priced 'net' cost, i.e. without overheads and profit being included in the rates. Overheads and profit are expressed as both a lump sum and a percentage for each contract.

Columns 9, 10 and 11 contain the net cost of labour, plant and materials as calculated for the tender or as revised, if the revisions have been agreed with the client. Column 9 probably represents the greatest element of risk in the contract.

To ascertain how the works are progressing at any time during the course of the contract's life it is necessary to make a review. These reviews are usually made monthly so that the time and cost of carrying out the review may be offset to some extent by using the data as the basis of interim applications for payments on account.

Figure 7.10 shows a pro forma which may be used to record the interim review data. The review uses figures derived from Figs. 7.8 and 7.9.

Interpretation of data

The data contained in columns 1–4 is taken from the contract data analysis but the contract sum and the contract period may have changed from those contained in the original tender.

Columns 5–11 contain data regarding standards and performance during the whole period of the contract up to the time of the review. Contract A, the Wellington Road Housing Contract, is in the seventeenth period (month) which should, as can be seen from column 4 in Fig. 7.10, have been the last month for this contract.

It is apparent from column 6 that the budget sum £825,500 is also the contract sum. In fact, the value of work which has been carried out is only £755,820; £69,680 less than the budget sum. The value of work included in column 7 is the sum which the quantity surveyor would be prepared to recommend for

Interim Review Data for Period October 1982 ......   Sheet 1 ....

1	2	3	4	5	Cumulative results						Current period results						18
					6	7	8	9	10	11	12	13	14	15	16	17	
Contract	Code ref.	Contract sum	Contract period (mnths)	Period no.	Budget sum	Work value	Cost	Profit (dif. cols. 7/8)	Loss (dif. cols. 7/8)	Profit/loss as %age	Budget sum (£)	Work value	Cost	Profit (dif. cols. 13/14)	Loss (dif. cols. 13/14)	Profit/loss as %age	Remarks
Wellington Rd. Housing	WRH 80/8	825,500	17	17	825,000	755,820	727,600	28,220	—	3.88	34,600	32,400	29,800	2,600	—	8.02	3-week extension for inclement weather
Allbourne Social Cent	ASS 80/9	75,000	10	10	75,000	76,550	66,920	9,630	—	14.39	5,500	5,800	5,100	700	—	13.73	
Folton Comp Sch	FCS 81/11	1,200,700	18	9	758,700	702,100	704,300	—	2,200	(0.31)	96,600	79,900	77,300	2,600	—	3.36	
Ashley Pl'ce Hotel	APH 81/12	982,000	18	7	441,800	446,400	409,000	37,400	—	9.14	72,200	73,400	67,600	5,800	—	8.58	

Fig. 7.10 Interim review data

payment to the sub-contractor, subject to deductions for retention, etc. and subject to any adjustments which the sub-contractor may consider it necessary to make to ensure that when comparing the budget sum and the value the comparison was of like with like.

The sub-contractor (or more probably his surveyor) would, for instance, make adjustment for variation of price (increased costs due to inflation), if the independent quantity surveyor had included these in his valuation but the budget sum did not include such increases.

Similarly, if the quantity surveyor had, in the opinion of the sub-contractor, undervalued or overvalued the work carried out to date, he would make appropriate adjustments to ensure that a valid comparison was being made. It is by no means uncommon for a contractor to endeavour to boost (overvalue) the value of his work to obtain as large an interim payment on account as possible and so improve the cash-flow. If he succeeds, good luck to him, but he would be deceiving himself if he used an inflated valuation when making financial reviews.

Similar comments apply to column 8: the cost. This should be adjusted to take into account times, which, for instance, are not yet entered into the prime cost ledger. Columns 9 and 10 are calculated by taking the difference between columns 7 and 8.

Column 11 is calculated by taking

$$\frac{\text{col. 9 or col. 10} \times 100}{\text{col. 8}}$$

expressed as a percentage. As there is only one column for this percentage, a negative percentage can be indicated by placing it in brackets – the normal accounting practice.

The figures in columns 12–17 are obtained in a similar manner to those in columns 6–11, but they represent the work carried out during the month under review.

It is important to note that the budget sum in column 12 is the sum which should have been carried out during period 14 (see Fig. 7.11). This is because the contract is running behind schedule.

Figure 7.11 is the budget plan for one of the sub-contractor's contracts, the Wellington Road Housing contract. The plan allocates the sums included in the estimate to the period, the appropriate month in this instance, when it is planned to carry out the work.

Ascertaining variances: profit, loss and progress
The data discussed above may be used to identify variances between expectation and achievement in terms of finance and progress. When examining each contract, the controller should ask four questions:
(a) are we making a profit,
    (i) overall? (the cumulative results are the indicator)
    (ii) this month? (the current period results are the indicators);
(b) is our progress satisfactory,

Contract Budget Plan for Wellington Road housing.... Sheet 1......

Code ref. WRH/80/8 Client Pilborough DC ~~FMG~~ / VOP Architect PDC own QS PDC own

Form of Contract SFBC/LA Start date APR 81 Contract Period 17months Damages £300/week

1	2		3		4		5		6		7		8		9	
Period (monthly)	Contract sum		Over'hd & pr'ft		Labour		Plant		Materials		Preliminaries		Own sub-con		Noms-c's+sups	
	Current	Cuml.	Current	Cuml.	Current	Cuml.	Current	Cuml.	Current	Cuml.	Current	Cuml.	Current	Cuml.	Current	Cuml.
1 April 81	18,190	18,190	1,140	1,140	1,200	1,200	300	300	2,800	2,800	10,900	10,900	1,850	1,850	—	—
2 May	20,050	38,240	1,260	2,400	1,650	2,850	700	1,000	5,200	8000	6,220	17,120	2,920	4,770	2,100	2,100
3 June	26,240	64,480	1,650	4,050	2200	5,050	1,450	2,450	5,900	13,900	6,950	24,070	3,650	8,420	4,440	6,540
4 July	33,120	97,600	2,080	6,130	3,600	8,650	1,450	3,900	6,700	20,600	7,300	31,370	5,760	14,180	6,230	12,770
5 August	40,920	138,520	2,570	8,700	4,300	12,950	1,750	5,650	8,350	28,950	7,300	38,670	7,830	22,010	8,820	21,590
14 July (82)	34,670	755,820	2,320	47,460	2,200	62,400	1200	23,600	5,760	108,550	6,800	106,850	8,970	190,850	7,950	216,110
15 August	28,780	784,600	1,810	49,270	1,800	64,200	900	24,500	4,650	113,200	6,450	113,300	8,300	199,150	4,870	220,980
16 September	23,790	808,390	1,490	50,760	1,200	65,400	700	25,200	3,400	116,600	6,200	119,500	7,900	207,050	2,900	223,880
17 October	17,110	825,500	4,840	55,600	600	66,000	400	25,600	1,800	118,400	4,300	123,800	3,550	210,600	1,620	225,500

Fig. 7.11 Contract budget plan

(i)   overall?

(ii)   this month?;

(c)   what are the trends for completion of the contract,

    (i)   in terms of progress?

    (ii)   financially?;

(d)   what, if any, action should be taken?

    These questions may be asked for each of the contracts shown in Fig. 7.10, and unless otherwise stated, the columns, percentages, amounts, etc. referred to in the following questions are taken from Fig. 7.10.

Wellington Road Housing Contract

Q.1(a) *From column 11 we can see that we are making a profit overall. But at 3.88 per cent, it is lower than our estimated profit of 7.22 per cent, as shown in Fig. 7.9*

Q.1(b) *Our profit for the month's work is above the estimated margin (8.02 per cent compared with 7.22 per cent). In practice, it is difficult to make precise measurements of value and cost as there are numerous factors which affect them. Precision is not, however, essential as we are interested in trends and the action which should be taken.*

Q.2(a) *Progress can be measured by comparing the budget sum with the work value (cols 6 and 7). The difference between columns 6 and 7, £69,680, represents approximately 3 months work. And as the budget sum is in excess of the work value, clearly the contract is running behind schedule.*

Q2.(b) *Comparison of columns 12 and 13 indicates that the current period has continued the disappointing progress demonstrated by the contract overall. It is unusual for time to be recovered during the final month or two of a contract's life. It is some comfort that the work is being carried out profitably, even if the value of work is less than planned.*

Q.3 *Progress is considered first here, because until it has been assessed, the financial trend cannot be calculated.*

Q.3(a) *Column 2 of Fig. 7.11 indicates that at least three months will be required to complete the contract, probably four.*

Q.3(b) *It can be seen in column 7 of Fig. 7.11 that the monthly preliminaries budget is between £6,000 and £7,000, reducing in the final month. The indications are, therefore, that approximately £20,000 will be expended on preliminary costs above and beyond the budget.*

The overhead and profit margin stands at £28,220, and if the current period results (col. 15) continue for the remainder of the contract, the margin could improve to, say, £2,600 times 3, approximately £8,000. However, it would be folly to bank on that trend continuing.

The final margin will probably be £28,220 plus, say, £5,000 (margin on remainder of work) – £20,000 (preliminary cost on over-run) equalling approximately £13,000.

From this margin one should make allowance for liquidated damages to which the client may be entitled. It is known that a 3-week extension of time has been granted and more may be forthcoming. Nevertheless, the contractor may be liable for up to three months damages. At £300 per week his margin might be further reduced by, say, 13 (weeks) × £300 = £3,900.

Q.4 *It is too late in the final month of a contract to initiate effective remedial action. The group may well have taken steps earlier in the contract's life to retrieve a more serious situation. There are numerous possible causes of the adverse variances. Identifying them is the first step towards corrective action; that will be discussed later.*

Allbourne Social Centre

Q.1(a) *Columns 7 and 8 show that extra work has been absorbed in the contract. The profit is in excess of the estimate overall.*

Q.1(b) *The current period result is similar to the overall result.*

Q.2(a) and (b) *The contract is complete on programme.*

Q.3 *This is not relevant as the contract is complete.*

Q.4 *No action appears to be necessary.*

Folton Comprehensive School

Q.1(a) *Comparison of columns 7 and 8 shows the work value to be less than the cost; the contract is sustaining an overall loss. The loss is not great but at this stage, the contract should be showing a 'profit' of approximately £37,000 (5.2 per cent of £704,300).*

Q.1(b) *The current period results appear to be marginally more satisfactory. The profit margin is less than planned, and one would need to consider more than one month's results before a trend could be established, but at least the adverse trend apparent from the 'cumulative result' has been reduced this month.*

Q.2(a) *The budget sum in column 6 is £758,700 and the work value, column 7, is £702,100; a difference of £56,600. In practice, the group would have a budget plan for this contract, in the format as in Fig. 7.11, from which the budget sums for each month could be ascertained.*

However, from column 12 we can see that this month's budget value is £98,600. It follows that an approximation of delay would be

$$\frac{56,600}{98,600} \text{ months}$$

Assuming a twenty-two-day month, this represents thirteen days or two-and-a-half weeks delay.

Q.2(b) *A similar comparison of columns 12 and 13 shows a delay for the month of*

$$\frac{(98,600 - 79,900)}{79,900} \times 22 = 5 \text{ days delay.}$$

Q.3(a) *A comparison of this month's variance with the previous month would show if the present trend is persisting. Assuming this month's variance is typical and that it was likely to continue until the end of the contract period, the total delay would be*

$$\frac{5}{22} \ (days) \times 9 \ (months - remaining\ period) = 2\ months\ delay.$$

Q.3(b) *An approximation of the monthly cost of preliminaries would be*

$$\frac{£145,800}{18\ (months)} \ \ (see\ Fig.\ 7.9) = £8,100$$

If the prediction of two months made above is correct, the loss on preliminary costs would be £16,200. It would be foolish and incorrect to extrapolate the loss indicated in Q.1(a) above for the remainder of the contract period. But the indications are that further loss, probably substantial, will be sustained.

Q.4 *The management group would no doubt look closely at the results, and instruct the manager for this contract to undertake detailed investigations if this action had not already been taken as the result of previous reviews.*

Ashley Place Hotel   This contract is different from the last in that it is ahead of plan and its profit level is in excess of anticipation. It is however, still early days. It is not necessary to make an arithmetical analysis of this contract as the method is similar to that used for the previous contract. The management group would, however, pay close attention to this contract as it is at the stage when steps can be taken to keep it on the right path.

The four contracts examined above demonstrate the most usual trends which contracts follow. The analyses of the contracts show how the information can be read to determine not only financial trends but also progress trends. Indeed, experience suggests that budgets indicate progress better than a supposedly analytical study of progress charts. It is remarkably easy to read into progress charts whatever the reader wishes to see. The firm's actual income from the contract is often a much better guide.

What format should the management group adopt for its review? Procedures for reviews vary from firm to firm. Some firms, often the larger ones, tend to call the managers and surveyors for each contract into the office once a month and conduct formal reviews into all the contracts in quick succession. Others carry out their reviews on site, the contracts director maintaining communication with the sites by means of personal visits.

The need to carry out the monthly review provides the discipline he needs to make these visits. He feels, too, that there are some aspects of a contract's performance which are not apparent from reports and he likes to see for himself.

The management group would have more information about the contracts than has been provided above. There would be budget plans for each contract, and members of the management group would have previous monthly review data to refer to should they need to do so.

Furthermore, they would know the action taken previously. The review is management by exception.

Contracts which can be seen from columns 6–11 of Fig. 7.10 to be on course would be given less attention than those which were not.

The data contained in the pro forma used by the management group is insufficient for it to decide upon a detailed course of action. But then that would not be its function. Having identified exceptions it would refer the whole issue to the appropriate contract manager and/or contract surveyor for action and report.

The contract manager and surveyor would have the information from which the pro forma discussed above were derived.

Budgetary planning and control for the company

The principles of budgetary planning and control described above for the planning and control of individual projects may be applied to the company as a whole through a system of responsibility accounting.

Responsibility Accounting

Responsibility accounting is a system of management which attempts to provide the entrepreneur 'at the top' of an organization with a means of managing the whole economic task of an enterprise. It provides a framework with 'levels of responsibility' each level of which has a budget which is controlled by an appropriate 'controller'. Figure 7.12 shows a typical organization structure for a medium/large specialist contracting organization. It is important that the *structure* of the organization is shaped to facilitate the provision of sub-goals which relate to the company's objectives.

At Level 1, the managing director controls a series of divisions, (a)–(g) in Table 7.2, which are 'profit' or 'cost centres'.

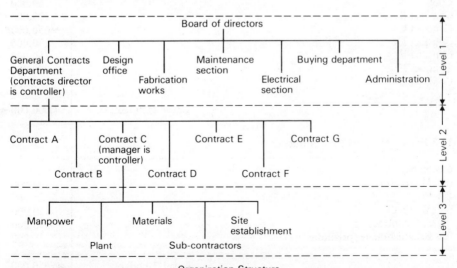

Organization Structure
(showing levels of control within General Contracts Department)

Fig. 7.12 Organization structure

Each division has a target or budget as shown in that figure. At level 2 the contracts director controls contracts A–G which are his sub-goals; profit centres, and so on to level 3.

The number of levels in the organization may increase as the size or complexity of the company increases but it is important that the organization structure is shaped to its function, not vice versa.

Figure 7.13 shows a monthly profit/loss statement for the controller at level

Table 7.2 Budgetary Responsibility

LEVEL 1 – Board of Directors

	Budgets (£s)
Controlling:	
(a) General Contracts Department	4,000,000 ◄
(b) Design office	500,000
(c) Fabrication plant	150,000
(d) Maintenance section	70,000
(e) Electrical section	200,000
(f) Buying Department	30,000
(g) Administration Department	60,000
Level 1 budget:	£5,010,000

LEVEL 2 – Contracts Director

	Budgets (£s)
Controlling:	
Contract A	750,000
Contract B	575,000
Contract C	► 650,000
Contract D	150,000
Contract E	350,000
Contract F	950,000
Contract G	575,000
Level 2 budget:	£4,000,000 ◄

LEVEL 3 – Manager for Contract C

	Budgets (£s)
Controlling:	
(a) Manpower	100,000
(b) Plant	25,000
(c) Materials	150,000
(d) Sub-contractors	350,000
(e) Establishment (preliminaries)	25,000
Level 3 budget:	► £650,000

1. *Column 2* in the statement shows the *planned turnover* (budget) for each profit/cost centre for the month. In the case of the general contracts department this figure would be the sum of all the contracts.

Column 3 shows the *value* of the work (or service) carried out during the month.

Column 4 shows the *actual cost* of the work.

Column 5 shows the sum which the contractor received after deduction of retention sums, etc.

Columns 6 and 7 represent the difference between columns 3 and 4, the profit or loss for each cost centre for the month.

Column 8 states the profit or loss as a percentage (col. 6 or 7 × 100).

The cost of the buying department and administration, (f) and (g), may be treated as part of the establishment cost (see Ch. 4) or the cost may be allocated to profit centres (a)–(e) as direct charges. Whichever alternative is adopted when *charging* the costs of (f) and (g), it is important that these 'service' departments have budgets allocated to them.

Planning and control 'rules'

Planning and control is a part of the sub-contractor's management system which lends itself to computerization which should greatly reduce the clerical chores

Monthly profit/loss statement for month .JUNE.82.							
1. Profit/cost centre	2. Budget (£)	3. Actual wk/serv. value (£)	4. Actual cost (£)	5. Sales (£)	6. Profit (£)	7. Loss (£)	8. Profit/ loss (as %age)*
(a) General Contracts Department	380,000	405,000	370,000	385,000	35,000	–	9.46
(b) Design office	42,000	36,000	31,000	30,000	5,000	–	16.13
(c) Fabrication works	12,500	9,000	8,000	6,000	1,000	–	12.50
(d) Maintenance section	6,000	8,000	7,000	4,000	1,000	–	14.29
(e) Electrical section	17,000	13,000	11,000	12,000	2,000	–	18.18
(f) Buying Department	2,500	–	3,000	–	–	3,000	–
(g) Administration	5,000	–	5,500	–	–	5,500	–
Totals	465,000	471,000	435,500	437,000	44,000	8,500	8.15
Monthly profit/loss statement at level 1			–	435,500	8,500	–	*Col. 6 or 7 Col. 4
			Cash Flow	+ 1,500	35,500	Profit	× 100

Fig. 7.13 Information requirement for monthly control

133

involved and increase the speed with which essential data is available for the manager. The most mechanized systems are, however, dependent on input from human agencies to at least some extent. The site supervisor who fails to return delivery vouchers for materials to head office may, for example, lead to management gaining a false impression of project profitability – until invoices for the unreported materials arrive!

An important aspect of planning and control is, therefore, acknowledgement of the human factor and involvement of the personnel upon whom the implementation of the system depends. If personnel are actively concerned in *preparing* the plan or control system they will have a vested interest in their success. If they are not, they may find every difficulty in implementation just to prove that the people who were sufficiently rash to plan, as it were, unilaterally, are out of touch with the practicalities of life.

It is widely accepted that planning is not an exact science; numerous factors militate against success. The difficulties of assessing the project duration and cost are, for example, not restricted to the construction industry, Various 'rules' state:

Westheimer's Rule. 'To estimate the time it takes to do a task; estimate the time you think it should take, multiply by two, and change the unit of measure to the next highest unit. Thus allocate two days for a one-hour task' (PD–OR).

Murphy's Law. Everything takes longer than you expect (PD–OR).

Finagle's Delay Formula. After adding two weeks to the schedule for unexpected delays, add two more for the unexpected unexpected delays (PD–OR).

Cheops' Law. Nothing ever gets built on schedule or within budget (PD–OR).

It is accepted that 'If anything can go wrong it will' (*Chisholm Effect* (PD–OR)) and 'If on an actuarial basis there is a 50:50 chance that something will go wrong, it will actually go wrong nine times out of ten' (*Economist's Law*, PD–OR).

The importance of regularity and frequency when exercising control, referred to earlier in the chapter, is endorsed by *Golub's Third Law*, that 'the effort required to correct course increases geometrically with time' (PD–OR).

The rules should be taken with a pinch of salt although sub-contractors on building projects may take some comfort from Gerald Weinberg, computer scientist, University of Nebraska whose law reads: 'If builders built buildings the way programmers wrote programmes, then the first woodpecker that came along would destroy civilization.'

Management by exception

The methods of budgetary planning and control which are discussed in this chapter demonstrate the principle of *management by exception*. This approach accepts that there are time and resource constraints which prevent the manager from devoting too great a part of his time to monitoring and controlling work in progress. He must therefore decide which parts of the work in progress should be given priority. Adopting the exception principle, those parts of the works, projects, etc. which are in some way exceptional become his first concern. In the examples used above it is those projects which are outside of their budgets or on which the actual progress does not match programmed progress.

Management by crisis

The *management by crisis* approach requires the manager to concentrate on the current crisis or crises. He is responding to events rather than organizing and controlling them. It is sometimes argued that this approach is not 'management' and it is appreciated that there is some justification for this argument but it must also be appreciated that however carefully plans are prepared there are invariably factors beyond the manager's control which will make some crises inevitable.

Perhaps a measure of managerial success in regard to planning, organizing and controlling, is the amount of time which the manager spends managing by objectives, by exception and by crisis. In an ideal world the first-mentioned would occupy the majority of his time. Experience shows that in the building industry, the manager spends more of his time dealing with day-to-day crises than he would wish.

Event times for operations between Nodes 59 to 75.

Node	earliest event time	latest event time
59	238	254
63	246	262
69	242	262
73	244	262
75	246	—

Chapter 8

Insurances

Having prepared the plan and arranged to monitor and control progress and cost the specialist contractor is in a position to commence the works. Before doing so, however, it is necessary for him to consider his obligations under the contract and for his own needs in respect of insurances.

There is often considerable confusion over the question of what the sub-contractor is responsible for under the terms of his sub-contract and even more confusion as to what he needs to insure. The problem often arises due to the fact that the sub-contractor believes, quite incorrectly, that there is a duplication of cover with that provided by the main contractor. In this chapter we will therefore very briefly look at the position under the main contract, followed by a more detailed analysis of the sub-contracts.

Position under the main contract conditions

Clause 20.1 – the contractor indemnifies the employer against death or bodily injury

Clause 20.2 – the contractor indemnifies the employer against loss of or damage to property (other than caused by a peril accepted by the employer under clause 22B or 22C)

Clause 21.1 – the contractor is required to insure against his clause 20 liabilities

Clause 21.2 – the contractor is required to arrange cover for the benefit of the employer where required by provisional sum item in the bills. It should be noted that this clause has no equivalent in the sub-contracts as once the cover is arranged the policy will provide cover for the employer whether the loss or damage arises out of the main contractor's or the sub-contractor's work.

Clause 22 A – the contractor is required to insure for the benefit of the employer and himself, loss or damage to the works and materials for incorporation therein. The contractor is responsible for carrying out and completing the works (see art. 1 of the articles of agreement and cl. 2.1 of the main contract) and is therefore, in effect, responsible for any loss or damage to the works (unless he can

136

prove that he is relieved from such responsibility elsewhere in the contract). However, under the terms of this clause the main contractor is only required to insure for 'the Clause 22 perils' as defined in clause 1.3:

Clause 22 Perils:

fire, lightning, explosion, storm, tempest, flood, bursting or overflowing of water tanks, apparatus or pipes, earthquake, aircraft and other aerial devices or articles dropped therefrom, riot and civil commotion, (excluding any loss or damage caused by ionizing radiations or contamination by radioactivity from any nuclear fuel or from any nuclear waste from the combustion of nuclear fuel, radioactive toxic explosive or other hazardous properties of any explosive nuclear assembly or nuclear component thereof, pressure waves caused by aircraft or other aerial devices travelling at sonic or supersonic speeds).

Clauses 22B and 22C are alternative to 22A where the contractor is relieved of responsibility for the 'Clause 22 Perils' and therefore also for insuring the works. In so far as clause 22B is concerned this is because the employer, for whatever reason, wishes to retain responsibility for those perils whether or not he decides to insure them. In so far as 22C is concerned this is due to the fact that the contract is for alterations or extensions and the employer will therefore extend his insurance on the existing buildings to cover these perils to the works. The contractor is, however, responsible for any other loss or damage to the works other than by 'Clause 22 perils'.

When the sub-contractor is familiar with the main contract requirements he is in a position to consider the equivalent clauses under the sub-contracts and how these relate to the above. The comments that follow apply equally to sub-contract NSC/4 or DOM/1, i.e. the nominated sub-contract and the domestic sub-contract respectively.

Position under the sub-contract

Incorporated provisions liability

The sub-contractor's liability under incorporated provisions of the main contract are contained in clause 5:

5.1 The sub-contractor shall:

5.1 .1 observe, perform and comply with all the provisions of the Main Contract as described by or referred to in the Tender, Schedule 1 on the part of the Contractor to be observed, performed and complied with so far as they relate and apply to the Sub-Contract Works (or any portion of the same). Without prejudice to the generality of the foregoing, the Sub-Contractor shall observe, perform and comply with the following provisions of the Main Contract Conditions: clauses 6,7,9,16,32,33 and 34; and

5.1 .2 indemnify and save harmless the Contractor against and from:

.2 .1 any breach, non-observance or non-performance by the Sub-Contractor or

137

his servants or agents of any of the provisions of the Main Contract referred to in clause 5.1.1; and

.2 .2 any act or omission of the Sub-Contractor or his servants or agents which involves the Contractor in any liability to the Employer under the provisions of the Main Contract referred to in clause 5.1.1.

5.2 Nothing contained in the Sub-Contract Documents shall be construed so as to impose any liability on the Sub-Contractor in respect of any act or omission or default on the part of the Employer, the Contractor, his other sub-contractors or their respective servants or agents nor (except by way of and in the terms of the Agreement NSC/2) create any privity of contract between the Sub-Contractor and the Employer or any other sub-contractor.

Clause 5 has no real equivalent in the main contract. It requires the sub-contractor to observe and comply with all the provisions of the main contract and to indemnify the contractor in breach, etc. by the sub-contractor of the provisions of the main contract and any act or omission of the sub-contractor which involves the contractor in any liability to the employer under the main contract.

It can be seen from this clause that it endeavours, amongst other things, to make it clear that the contractor will *not* be responsible for any liabilities that fall upon him as a result of the actions of the sub-contractor. This goes some way to explaining how it is that both the sub-contractor and the contractor need their own independent insurance covers.

Injury to persons and property – indemnity to contractor

The sub-contractor's liabilities are contained in clause 6:

6.1 The Sub-Contractor shall be liable for and shall indemnify the Contractor against any expense, liability, loss, claim or proceedings whatsoever arising under any statute or at common law in respect of personal injury to or the death of any person whomsoever arising out of or in the course of or caused by the carrying out of the Sub-Contract Works unless due to any act or neglect of the Contractor or his servants or agents or of any other sub-contractor of the Contractor engaged upon the Main Contract Works or any part thereof his servants or agents or of the Employer or of any person for whom the Employer is responsible.

Clause 6.1 is really the direct comparison with clause 20.1 of the main contract and requires the sub-contractor to indemnify the contractor in respect of death or injury to any person whether they be an employee or third party to the sub-contractor or the contractor, as the case may be.

The indemnity is extremely wide in that the sub-contractor accepts responsibility not only for his own negligence but in respect of *any* death or injury, howsoever caused. The only escape for the sub-contractor is if he can prove, and the onus of proof is on the sub-contractor, that the death or injury was due to any act or neglect of the contractor, his servants or agents, other sub-contractors to the contractor, their servants or agents, the employer, or any person for whom the employer is responsible. It is quite clear this clause goes beyond the sub-contractor's normal liability at English common law. For this reason it is essential that the sub-contractor has adequate cover under his employer's liability and public liability policies to cover liability assumed under contract or agreement.

6.2   Except for such loss or damage as is referred to in clause 8.1.1.1, or is at the sole risk of the Employer under the relevant and applicable provisions of the Main Contract Conditions, the Sub-Contractor shall be liable for and shall indemnify the Contractor against any expense, liability, loss, claim or proceedings in respect of any injury or damage whatsoever to any property real or personal in so far as such injury or damage arises out of or in the course of or by reason of the carrying out of the Sub-Contract Works and provided always that the same is due to any negligence, omission or default of the Sub-Contractor, his servants or agents.

Clause 6.2 is the direct comparison with clause 20.2 of the main contract. Under the clause the sub-contractor indemnifies the contractor in respect of injury or damage to any property provided it arises out of the contract works and also provided that it is due to the negligence, omission or default of the sub-contractor or his servants or agents. This clause is therefore not as wide in its application as clause 6.1 and probably does not extend the sub-contractor's common law liability. The clause does, however, make special provision for the exclusion of loss or damage which is at the sole risk of the contractor under clause 8.1. This is to cater for the position where the main contract requires the contractor to insure the works including sub-contract works against the 'Clause 22 Perils'. It should be noted, however, that the sub-contractor is not relieved of responsibility totally for loss of or damage to his works but only in respect of those perils referred to in clause 22 of the main contract. It will be noted from clause 8.1.1.2 that if the contractor has insured the 'Clause 22 Perils' the sub-contractor shall not be responsible for any damage howsoever caused nor be under any obligation to affect any insurance.

Similarly in clause 8.1.2.2 if the employer has accepted responsibility for the clause 22 perils the sub-contractor shall not be responsible for any loss or damage nor required to effect any insurance as in these circumstances the works are at the 'sole risk of the Employer' under clause 8.2.1.

Personal injury and injury or damage to property

The sub-contractor's liabilities in these respects are contained in clause 7:

7.1   Without prejudice to his liability to indemnify the Contractor under clauses 5 and 6 the Sub-Contractor shall maintain such insurances as are necessary to cover the liability of the Sub-Contractor in respect of personal injury or death arising out of or in the course of or caused by the carrying out of the Sub-Contract Works not due to any act or neglect as referred to in clause 6.1 and such insurances as may have been detailed in the Tender, Schedule 2, item 4 to cover the liability of the Sub-Contractor in respect of injury or damage to property real or personal arising out of or in the course of or by reason of the carrying out of the Sub-Contract Works and caused by any negligence, omission or default of the Sub-Contractor, his servants or agents.

7.2　The insurance in respect of claims for personal injury to, or the death of, any person under a contract of service or apprenticeship with the Sub-Contractor, and arising out of and in the course of such person's employment, shall comply with the Employer's Liability (Compulsory Insurance) Act 1969, and any statutory orders made thereunder or any amendment or re-enactment thereof. For all other claims to which clause 6.2 applies the insurance cover shall be the sum stated in the Tender, Schedule 1, item 10 (or such greater sum as the Sub-Contractor may choose) for any one occurrence or series of occurrences arising out of one event.

Clause 7 is the direct comparison with clause 21 of the main contract and therefore requires the sub-contractor to insure against his liabilities (as distinct from the liability of the contractor) as referred to in clause 6.

The clause does not state specifically what policies are to be negotiated or how they are to be arranged but only to arrange 'such insurances as are necessary to cover the liability, etc.' In the main, the policies to be arranged are an employer's liability policy and a public liability policy (sometimes referred to as a Third Party policy). In some circumstances these are issued as one single document which may then be known as an indemnity policy.

The clause stipulates that with regard to death or injury to employees the policies shall comply with the Employers' Liability (Compulsory Insurance) Act 1969 and that for 'other claims' shall be not less than the sum stated in the tender document in respect of any one occurrence or series of occurrences arising out of any one event. The Employers' Liability (Compulsory Insurance) Act has various stipulations but the major one for the purposes of this chapter is that it requires a minimum limit of indemnity in respect of death or injury to employees of £2 million arising out of any one incident. It is fair to say that no British insurance company would issue an employer's liability policy in circumstances where the Act applied that did not have at least a limit of £2 million and it is far more normal to find that this policy will have no financial limit whatsoever. There are, of course, other limitations and conditions which must be watched carefully.

With regard to 'other claims' these would normally be the subject of a claim against a public liability policy. The limit quoted in the tender, it should be remembered, is only a *minimum* limit which the employer will accept and in no way limits the sub-contractor's liability to this figure. In any case this policy will also have to provide for claims from members of the public who are not a party to the contract and it is quite clear that they could not in any case be a party to a restricted limit of indemnity, even if it were such, which it is not. The sub-contractor should therefore consider the limit of indemnity he wishes to insure quite independently of what he is required to insure by this clause, subject to the minimum quoted in the tender.

'Clause 22 Perils'

The sub-contractor's liabilities in respect to 'Clause 22 Perils' are contained in clause 8:

8.1 .1 .1   Where clause 22A of the Main Contract Conditions applies the Contractor shall effect the insurance required by clause 22A, namely to insure against loss and damage by the 'Clause 22 Perils' all work executed and all unfixed materials and goods, delivered to, placed on or adjacent to the Works and intended for incorporation therein but excluding temporary buildings, plant, tools and equipment owned or hired by the Contractor or any sub-contractor and keep such work, materials and goods so insured until (subject to clause 18.1.4 of the Main Contract Conditions) the date of issue of the Certificate of Practical Completion under clause 17.1 of the Main Contract Conditions.

    .1 .2   The Sub-Contractor shall not be responsible for any loss or damage however caused where clause 22A of the Main Contract Conditions applies nor be under any obligation to effect any insurance in respect of the 'Clause 22 Perils'.

8.1 .2 .1   Where clause 22B or clause 22C of the Main Contract Conditions applies then all work executed and all unfixed materials and goods, delivered to, placed on or adjacent to the Works and intended for incorporation therein (except temporary buildings, plant, tools and equipment owned or hired by the Contractor or any sub-contractor) and, where clause 22C of the Main Contract Conditions applies, the existing structures are at the sole risk of the Employer as regards loss or damage by the 'Clause 22 Perils'.

    .2 .2   Neither the Contractor nor the Sub-Contractor shall be responsible for any loss or damage however caused to which clause 22B or clause 22C of the Main Contract Conditions applies nor shall either be under any obligation to effect any insurance in respect of the 'Clause 22 Perils' referred to in the aforesaid clause 22B or clause 22C.

Clause 8.1 sets out the position with regard to insurance of the sub-contract works. It should, however, be made clear that the sub-contractor is responsible for all loss or damage to his works (see art. 1 of the Articles of Sub-Contract Conditions and comments under cl. 22A of main contract) except to the extent that he can show that he is relieved of responsibility by the sub-contract conditions. Clause 8.1 is, in fact, one of the clauses which does give him some relief from responsibility. As has been said previously, where clause 22A of the main contract applies the 'Clause 22 perils' are the responsibility of the contractor however these are caused and irrespective of whether it is as a result of the sub-contractor's negligence.

Similarly, where clause 22B or 22C of the main contract applies the employer accepts 'sole' responsibility for the 'Clause 22 perils' and once again the sub-contractor is therefore relieved from responsibility including damage caused by one of these perils due to his own negligence.

It cannot, however, be too strongly emphasized that the sub-contractor is only relieved of responsibility for loss or damage caused by any of the 'Clause 22 perils' and that he is therefore responsible for any other loss or damage which may occur, e.g. due to subsidence, collapse, theft, vandalism, malicious damage, accidental damage, etc. It should also be noted that the contractor or the employer is only relieving the sub-contractor for loss or damage caused by a 'Clause 22 peril' in respect of the works and unfixed materials for incorporation therein as stated in the clause. It does not relieve the sub-contractor in respect of loss or damage to temporary buildings, plant, tools and equipment owned or hired by the sub-contractor. The clause is strangely silent (as in the main contract) with regard to the sub-contractor's temporary works.

It is not clear whether temporary works are within the phraseology of 'all works executed' and therefore at the risk of the contractor or the employer for the 'Clause 22 perils' or not. It is therefore safer to assume that these are *not* within his phraseology and are therefore also at the risk of the sub-contractor. The sub-contractor would therefore be prudent to insure against his responsibility for loss or damage to his own works, excluding any loss or damage which is the responsibility of the contractor or employer as a 'Clause 22 peril' within the meaning of the clause.

Such policy can also extend to provide full All Risks cover if required on temporary works, temporary buildings, plant, tools and equipment of the sub-contractor or hired by him. This cover is commonly known as All Risks Contract Works Insurance or alternatively Contractors' All Risks Insurance.

Clause 8.2 reads:

8.2 .1 Where clause 8.1.1 applies, then in the event of any loss or damage being caused by any one or more of the 'Clause 22 Perils' to the Sub-Contract Works and any of the material or goods of the Sub-Contractor delivered to, placed on or adjacent to the Sub-Contract Works and intended for incorporation therein the Contractor to the extent of such loss or damages shall pay to the Sub-Contractor the full value of the same, such value in the case of loss or damage to the Sub-Contract Works to be calculated as if the reinstatement of the loss or damage had been carried out in accordance with instructions of the Architect as to the expenditure of a provisional sum.

.2 Where clause 8.1.2 applies, then in the event of any loss or damage being caused by any one or more of the 'Clause 22 Perils' to the Sub-Contract Works and any of the materials or goods of the Sub-Contractor delivered to, placed on or adjacent to, placed on or adjacent to the Sub-Contract Works and intended for incorporation therein, the Sub-Contractor upon discovery of the loss or damage shall forthwith give notice in writing to the Contractor of the extent, nature and location thereof. The occurrence of such loss or damage shall be disregarded in computing any amounts payable to the Sub-Contractor under or by virtue of the Sub-Contract; and

.2 .1 Where clause 22B of the Main Contract Conditions applies the Sub-Contractor with due diligence shall restore work damaged, replace or repair any unfixed materials or goods which have been destroyed or injured, removed and dispose of any debris and proceed with the carrying out and completion of the Sub-Contract Works. The restoration of work damaged, the replacement and repair of any unfixed materials and goods and the removal and disposal of debris shall be treated as if they were a Variation required by an instruction of the Architect under clause 13.2 of the Main Contract Conditions;

.2 .2 Where clause 22C of the Main Contract Conditions applies then:
- if the employment of the Main Contractor is determined under clause 22C.2.2 of the Main Contract Conditions clause 31 shall apply as if the employment of the Main Contractor was determined under clause 28 of the Main Contract Conditions but subject to the exception in regard to the application of clause 28.2.2.6 of the Main Contract Conditions referred to in clause 22C.2.2.2 of the Main Contract Conditions;
- if the employment of the Contractor is not determined under clause 22C.2.2 of the Main Contract Conditions then the Sub-Contractor shall with due diligence reinstate or make good such loss or damage, and proceed with the carrying out and completion of the Sub-Contract Works and shall receive such share of the monies paid to the Contractor in accordance with clause 22C.2.3 of the Main Contract Conditions as may be properly attributable to the Sub-Contract Works.

Clause 8.2 provides for the claims procedure in the event of loss or damage occurring to the sub-contract works. Clause 8.2.1 states, in effect, that where clause 22A of the main contract applies and damage is caused by one of the 'Clause 22 Perils' to the sub-contract works or materials for incorporation therein then the cost of reinstatement of these works and materials shall be paid for by the contractor as if this were a provisional sum item carried out on the instructions of the architect.

Clause 8.2.2 applies where clause 22B or 22C of the main contract is invoked and the clause is split between these two clauses as follows:

Where clause 22B of the main contract applies (i.e. where the employer has accepted responsibility for the 'Clause 22 Perils' for the works, whether or not insured) the sub-contractor is required 'with due diligence' to restore the damaged work and the sub-contractor will be paid for such additional work as if no loss had, in fact, occurred, i.e. he will be paid in full for the work he carries out.

Where clause 22C of the main contract applies a similar position occurs except that the sub-contractor is required to 'with due diligence reinstate or make good such damage' but in so far as payment is concerned he shall 'receive such share of the money paid to the contractor in accordance with clause 22C.2.3 of the Main Contract Conditions as may be properly attributable to the Sub-Contract Works', In practice this should mean that the sub-contractor is paid in full by the contractor for the additional work that he is called up to carry out providing it results from the operation of a 'Clause 22 Peril'.

If the damage to the sub-contract works is caused by something other than a 'Clause 22 Peril' then it is for the sub-contractor to pay for the cost of reinstatement (subject to him not being able to find relief elsewhere in the contract) subject to the provision of clause 8.3.

8.3 Subject to clause 8.1: [e]

8.3 .1 The Sub-Contractor shall be responsible for loss of or damage to all materials or goods properly on site for incorporation in the Sub-Contract Works (except any such materials and goods as have been fully, finally and properly incorporated into the Main Contract Works) other than for any loss or damage due to any negligence, omission or default of the Contractor, his servants or agents, or any other sub-contractor of the Contractor engaged upon the Main Contract Works or any part thereof, his servants or agents or of the Employer or any person for whom the Employer is responsible.

8.3 .2 Where materials or goods have been fully finally and properly incorporated into the Main Contract Works before practical completion of the Sub-Contract Works the Contractor will be responsible for loss or damage to such materials and goods except for any loss or damage caused thereto by the Sub-Contractor, his servants or agent.

8.3 .3 On practical completion of the Sub-Contract Works the Contractor will be responsible for loss or damage to the Sub-Contract Works properly completed and handed over except for any loss or damage caused thereto by the Sub-Contractor, his servants or agents.

There is always considerable confusion as to when the sub-contract works are no longer the responsibility of the sub-contractor and such responsibility passes to the contractor or the employer. Clause 8.3 endeavours to define when the sub-contractor is no longer responsible for the sub-contract works (other than his defects liability responsibilities). Clause 8.3.1 deals with the responsibility for loose materials for incorporation in the works and makes it quite clear that these are the the responsibility of the sub-contractor until such time as they are 'fully, finally and properly incorporated' into the main contract works. The exceptions to this are where the loss or damage is due to a 'Clause 22 Peril' which is the responsibility of the contractor or employer under the terms of clause 8.1 or where the sub-contractor can prove that the loss or damage was due to the negligence, omission, or default on the part of the contractor, his servants or agents, other sub-contractors, the employer, or any persons for whom the employer is responsible. In other words the sub-contractor is responsible for any loss or damage to loose materials unless it is due to a 'Clause 22 Peril' or the negligence of any other party to the main and/or sub-contract.

Clause 8.3 then deals with the situation of materials 'fully, finally, or properly incorporated' but before practical completion of the sub-contract works. In these circumstances the contractor is responsible for any loss or damage whatsoever unless that damage is actually caused by the sub-contractor or his servants or agents.

Clause 8.3 then deals with the situation upon practical completion where the contractor is once again responsible except for loss or damage caused by the sub-contractor, his servants or agents.

The difficulty with this clause is endeavouring to interpret the intention of the words 'fully, finally and properly incorporated'. There are numerous examples of the different sides to the argument having extreme opposing views. Merely to illustrate the difficulties, and not an endeavour to define intention, the following examples are given:

Situation 1

A plumbing sub-contractor providing wash basins, pipes, fittings etc. to a block of offices finishes work one night having hung four wash basins each of which is in a different state of 'completion'. The first is fully fixed including all screws, taps and pipes. The second has been fully and properly secured to the wall but has no taps and pipes connected. The third has only half its full complement of screws fixing it to the wall and the fourth has all screws, taps and pipes completed but the wash basin in some way is defective. The question then follows which if any of the above basins are fully, finally and properly incorporated. It could be argued that none are fully, finally and properly incorporated until such time as the whole toilet/washing facility area is completed. It could be argued that only the basin which is correctly screwed to the wall with its taps and pipes connected is within the meaning of the clause. At the stretch of one's imagination it could be argued that even the wash basin which has only half its complement of screws is fully, finally, and properly incorporated in so far as the work that had been done is concerned and in the last example the fact that the defect in the wash basin was a latent defect could mean that this, too, was fully, finally and properly incorporated. However, there is little doubt that apart from the first example there would be many people who would argue that each of these circumstances would not constitute work which is fully, finally and properly incorporated. It is not intended here to attempt to interpret the meaning as only the courts will decide how the words shall be interpreted given a specific set of circumstances.

Situation 2

Perhaps a clearer example would be an electrical contractor whose contract is to provide amongst other things electrical wiring in a new building. The contract calls for the sub-contractor to provide the conduit prior to the plastering but for whatever reason to supply the cables after the plastering has been completed. The sub-contractor attends on site and installs the conduit and leaves site until such time as he is called back to install the cable. During this period somebody intentionally or accidentally fills up the conduit with material in such a way that the only way of completing the contract is to remove the original conduit (which has now been plastered over) and replace it. Was the conduit fully, finally and properly incorporated at this stage or would it only have become so at the time the cable had been passed through the conduit?

There are, of course, numerous other examples which could be quoted but the above are hopefully sufficient to prove that there are problems in interpretation with this clause. Perhaps it would be sensible for the sub-contractor to come to some form of agreement with the contractor prior to commencing the work as to how 'fully, finally and properly incorporated' shall be interpreted bearing in mind the peculiarities of the contract in question. This may sound a laborious task but it can save a lot of difficulty in the event of a claim, certainly, until such time as the intention of these words has been thoroughly thrashed out by the courts.

145

8.4 The Sub-Contractor shall observe and comply with the conditions contained in the policy of insurance of the Contractor or of the Employer, as the case may be, against loss or damage by any of the 'Clause 22 Perils'.

Clause 8.4 merely states that in view of the fact that the sub-contractor, in effect, has the benefit of the contractor's or the employer's policy he must comply with the policy conditions of that policy, in order that the contractor or employer is not debarred from claiming under his policy.

8.5 Nothing in clause 8 shall in any way modify the Sub-Contractor's obligations in regard to defects in the Sub-Contract Works as set out in clauses 14.3 and 14.4.

Clause 8.5 merely states that nothing in clause 8 shall relieve the sub-contractor of his defects liability responsibilities.

9.1 As and when reasonably required to do so by the other party the Contractor (whether in respect of insurance taken out in pursuance of the terms of the Tender, Schedule 2, item 4, or in order to comply with the provisions of the Main Contract as described by or referred to in the Tender, Schedule 1) and the Sub-Contractor shall produce (and shall cause any other person to produce) for inspection by the other, documentary evidence that the insurances are properly effected and maintained but on any occasion the parties hereto may (but not unreasonably or vexatiously) require to have produced for inspection the policy or policies and premium receipts in question.

9.2 Should the Sub-Contractor make default in insuring or in continuing to insure as provided in the Sub-Contract, the Contractor may himself insure against any risk with respect to which the default shall have occurred and may deduct a sum or sums equivalent to the amount paid or payable in respect of premiums from any monies due or to become due to the Sub-Contractor or such amount may be recoverable from the Sub-Contractor by the Contractor as a debt.

Clause 9 provides that the sub-contractor shall produce documentary evidence that he has complied with the insurance requirement of the contract. This need not necessarily be the policies themselves but the clause does allow for these to be produced on demand; although not unreasonably so. The clause also provides for the situation where the sub-contractor does not adequately insure and in such circumstances the contractor may insure on the sub-contractor's behalf and charge the premium to the sub-contractor by way of deduction from amounts due to him.

Sub-contractor's responsibility for his own plant, etc.

The sub-contractor's liability in respect of his own plant are contained in clause 10:

10.1 The plant, tools, equipment or other property belonging to or provided by the Sub-Contractor, his servants or agents and any materials or goods of the Sub-Contractor which are not properly on site for incorporation in the Sub-Contract Works shall be at the sole risk of the Sub-Contractor, and any loss or damage to the same or caused by the same shall, except for any loss or damage due to any

146

negligence, omission or default of the Contractor, his servants or agents or sub-contractor (other than the Sub-Contractor, his servants or agents), be the sole liability of the Sub-Contractor who shall indemnify the Contractor against any expense, liability, loss, claim or proceedings in respect thereof.

10.2 Any insurance against any loss of or damage to or caused by the aforesaid plant, tools, equipment, or other property or to materials or goods of the Sub-Contractor which are not properly on site for incorporation in the Sub-Contract Works, shall be the sole concern of the Sub-Contractor.

Clause 10 makes it clear that for any materials or goods on site which are not for incorporation into the sub-contract works and also the sub-contractor's own or hired-in plant, tools, equipment or other property shall be at the sole risk of the sub-contractor and sub-clause 10.2 makes it clear that the responsibility for insurance, if any, of such property is the responsibility of the sub-contractor.

Other risks

There are, of course, many other clauses in the contract which have some bearing on the insurance position but to list all of these would be outside of the scope of this book. The above are the major clauses with an effect on the insurance situation.

Insurance policies

By way of summary it may be useful to state that the main site insurances required (or recommended) as a result of these conditions are as follows.

Employers' liability

This policy is to cover sub-contractor's legal liability for death or injury to his own employees. The policy to be in accordance with the Employers' Liability (Compulsory Insurance) Act 1969. The policy should not exclude liability assumed under contract or at least include liability assumed under contract under the JCT forms in order to comply with the terms of that form. Liability is, however, assumed under contract in many other circumstances, e.g. the Contractors's Plant Association form, agreements for trespass of air space, overhanging of scaffold, etc. and it would be wise for this to be catered for also. The sub-contractor should also make certain that there are no offending exclusions in particular with regard to specific types of work that would allow insurers to exclude liability in respect of any operation that he is carrying out.

Public liability (third party) policy

This policy is to pay for claims in respect of death, injury, loss or damage to third parties (basically anybody who is not an employee of the sub-contractor).

The policy has to be for a minimum limit of indemnity as stated in schedule 1 to the tender but the sub-contractor should consider what indemnity limit he desires and treat the tender figure merely as a minimum acceptable to the employer. The policy should not exclude liability assumed under contract for exactly the same reasons as quoted under the heading of 'employers' liability' above. There are numerous pitfalls with this particular class of insurance and it has been suggested that the drafting of such a policy is the job of an insurance person with specific knowledge of the construction industry and that this is not therefore a policy which can be bought straight off the shelf of any insurance company.

Contractors' all-risks policy

The sub-contractor is not technically required to insure this under the terms of his contract but nevertheless would be wise so to do. He can exclude loss or damage by the 'Clause 22 Perils' where these are accepted by the employer or the contractor leaving only the balance of all risks to be paid for by his insurers. The policy can also cover his plant, tools, temporary buildings, etc. for full all-risks cover, subject to policy terms, should the sub-contractor so desire.

It should also be made clear that where the sub-contractor takes out this form of policy it is the intention of the policy to cover only 'the Sub-Contract Works' as referred to in the contract and will not extend to include loss or damage caused to the contractor's 'Works' or the employer's existing buildings or contents. If the sub-contractor does cause damage to this property then subject to there being a legal liability upon him to pay for the cost of restoration, this would be a claim to be made under the public liability policy.

Meetings

'If more than one-third of one's time is spent at meetings then something is wrong'
(Peter Drucker)

'**meeting**, assembly of people for entertainment etc.; assembly for worship' (COD)

The previous chapters have been devoted mainly to obtaining and planning the sub-contractor's work. He should by now be involved with the other members of the building team which will inevitably entail meetings of one sort or another. There is a tendency for managers to convene meetings as the solution for all their problems. Arranging a meeting is an easy course of action to take. Many people prefer to talk rather than perform other, more tedious, tasks and calling a meeting gives the impression of action even if, in fact, a meeting is an 'action-substitute', a time-consuming talk-shop.

It is important to have a clear, well-defined aim for calling a meeting. There should be a reason and an 'end product'. It should be remembered that every time a meeting is called those who attend are taken away from the tasks they would otherwise be performing and costs will be incurred.

Aims of meetings

For most practical purposes the principal aims of meetings may be listed as shown in Table 9.1 but a meeting may have more than one of the aims in mind.

Table 9.1 Aims of Meetings

  (i) giving information and/or instructions to others;
 (ii) exchanging information between all those present;
(iii) generating and exchanging ideas;
(iv) formulating for decision-taking;
 (v) marketing and sales promotions;
(vi) social purposes;
(vii) encouraging staff participation and involvement in company development.

Check-list for convenor

The prospective convenor of a meeting should ask himself the questions listed in Table 9.2 before he actually decides to call a meeting. British Institute of

Table 9.2 Check-list for meeting convenor

(a) have I a clear purpose in mind?;

(b) if so, is a meeting the best way of tackling the situation?;

(c) even if it is the *ideal* way, is there another, more economical, less time-consuming way of tackling the situation?;

(d) if a meeting is essential, who should attend?;

(e) should there be a record of the meeting? How?;

(f) should other persons, not attending, be informed of what happened or what was decided?;

(g) have I all the data I need in order to contribute to best effect?;

(h) have the other persons who will be attending sufficient information in advance of the meeting to enable *them* to contribute to best effect?;

(j) is the proposed venue equipped for audio/visual presentations which may be necessary?

Management, Management Check-list 13, provides a more comprehensive list of questions for various types of meeting.

It should be a simple matter for the manager to answer the questions in the check-list if he has the Aims of Meetings in front of him.

For example, when answering question (a) he may find that *none* of the aims seem to be appropriate. In this event he may decide against calling a meeting. Similarly, if he found that his aim, was 'giving information etc.', aim (i), he may decide that a memorandum to the people concerned was a more economical method of giving the information unless he wished to observe the receiver's reactions or to have a quick response; he would decide on his course of action while answering questions (b) and (c). If the aim of the meeting is in categories (ii) or (iii) the answer to question (d) may indicate a large gathering but if the aim were 'making decisions', aim (iv), the size of the meeting should be small rather than large. Experience suggests that major actions are rarely decided by more than four people and that the usefulness of any meeting is in inverse proportion to the attendance. A large meeting, appropriate for aims (ii) or (iii) may, however, be useful for generating and developing ideas on which to base decisions. The answer to question (e) will almost certainly be 'yes' if the aim of the meeting is decision-making. A record of the meeting is likely to be necessary if it was decided to restrict attendence at the meeting and the answer to question (f) would, therefore, probably be 'yes'.

The convenor of the meeting will need to be well informed if the meeting aims are in categories (i), (ii) and (iv) (see Q. (g)), and others attending should be well informed, probably by means of discussion papers circulated in advance of the meeting.

Types of meeting

Meetings vary in type from highly formal parliamentary, central and local government councils, boards, committees, sub-committees and working parties to trade and professional conferences, company 'board meetings' to comply with statutory requirements and meetings of directors or partners to decide policy. In addition, most partners, directors and managers of organizations with which the sub-contract manager is likely to be concerned usually hold regular meetings for the purpose of organizing, monitoring and controlling the progress of their projects and their profitability.

Site meetings are a well-established phenomenon in the construction industry which, some say, are wasteful of resources, ineffective and should be abolished. Sub-contractors and quantity surveyors probably waste more time than most other parties because they are *expected* to attend site meetings (Chappell 1980).

Given adequate and timely information the contracting team should be able to carry out and complete the works without the need for site meetings with the designers. The adoption of site meetings is a symptom of poor information flow from design team to those who have to construct. The use of site meetings, as a means of reporting progress, is also questionable. There are ways of reporting progress which do not involve between five and twenty-five people spending a minimum of half a day attending on site. It has been said that: 'the more time you spend in reporting on what you are doing, the less time you have to do anything. Stability is achieved when you spend all your time doing nothing but reporting on the nothing you are doing'.

Whilst report-making is unlikely to monopolize a manager's time it is possible for him to spend a disproportionate amount of his time on non-productive paperwork of report-making type.

Action for meetings

To illustrate action which might be necessary when making arrangements for and conducting meetings we will use as an example the management of a General Contracts Department in an organization of the type shown in Fig. 7.12 (p. 131). The general contracts department might be managed on a day-to-day basis by a small *management group* led by the contracts director who is satisfied that the group's tasks of obtaining work, organizing, monitoring and controlling progress of projects and ensuring positive cash-flow and profitability will best be served by implementing a series of regular meetings. There are three phases in the progression of meetings:
(a) pre-meeting arrangements;
(b) conducting the meeting;
(c) implementing decisions taken.

Agenda

Pre-meeting arrangements

Having established, above, that the aims of the proposed meeting/s must be clear and well defined, the matters to be discussed should be listed in an agenda. As a series of regular meetings is planned a standard agenda should be prepared.

The items on the agenda will be related to the group's tasks.

A typical agenda might read:

Item

1. Apologies for absence.
2. Minutes of last meeting.
3. Matters arising (from the minutes of the previous meeting).
4. Enquiries from prospective clients.
5. Tenders for submission during next week.
6. Any other business.

Meeting frequency

For practical purposes meetings are normally cyclical (weekly, monthly or annual). A regular day of the week and time of the day should be arranged for the meetings which members will 'keep clear' in their diaries. In the absence of pre-arranged meetings members must be given adequate notice.

The items in the above agenda would need attention at frequent intervals because dealing with enquiries from prospective clients and submitting tenders must be matters of some urgency. The meeting should, therefore, be held at weekly intervals. In addition to the items listed above the group would need to review the department's finances and the progress of projects, but these matters may not justify weekly consideration other than in exceptional circumstances. Once a month, however, the group's agenda might be extended by the addition of the following items:

Item

6. Financial review.
7. Project review.
8. Accounts for settlement.
9. Any other business.

Attendance at meetings

For reasons of economy the number of persons attending the meetings should be kept to a minimum consistent with efficiency. If the management group comprised the contracts director, the senior estimator, the company/departmental secretary, the senior contracts manager/engineer and the senior surveyor, it may be appropriate for the first three persons listed above to attend the weekly meetings but for the group to be extended to include the senior contracts manager/ engineer and the senior surveyor for the monthly meetings.

Conducting the meetings

The agenda provides the framework for the meeting.

Chairman

The contracts director would normally act as the chairman for the meetings in the example. His duties being:

(a) to guide the members through the items on the agenda and ensure that they do not digress or exceed the time available for the meeting;

(b) ensure that each item is debated adequately and that all members have a fair hearing;

(c) to decide the order in which members should speak;

(d) to stimulate discussion;

(e) to bring the members to a conclusion on each item which should be expressed as an unambiguous resolution. The resolution should identify the person/s who are to take action;

(f) ensure that the resolution is accurately and concisely recorded, unless it is decided that a record is not required.

The chairman of a management group such as that referred to above would seek the views of members of the group which would help to shape decisions but ultimately the responsibility for decision-making would rest with the chairman who, as contracts director, is accountable to the board of directors for his department's performance.

The decisions of democratically constituted councils, boards and committees such as those of local authorities or professional institutions are made on a majority vote basis and the committee has *corporate responsibility*. But for both democratic committees and groups where the chairman has ultimate responsibility, the chairman should be seen to be fair to all members. It is a difficult role, which leaves the chairman open to charges of being 'dictatorial' or 'indecisive' (frequently both at the same time, but by different members of the group. Meanwhile, he has probably regarded himself as the *servant* rather than the *master* of the group).

The members

The members should assist the chairman to accomplish his duties. All discussion should be addressed to the chairman or the meeting will deteriorate into a chat-shop. Only one member should speak at a time and members should be restricted as to the number of occasions on which they speak.

Minutes

The minutes should record:

(a) the date on which the meeting was held;

(b) those present;

(c) separate, numbered minutes of the matters discussed recording briefly:

    (i) subject

    (ii) resolution with action to be taken

(iii)   names of person/s to undertake action
(d)   date, time and place of next meeting;
(e)   distribution list.

Table 9.3 shows part of a typical format for the minutes for meetings of a management group such as that described above.

Table 9.3 Format for minutes of meeting

Minutes of meeting held on Friday 18th June 19 . . . .

Those present were: P. Able (Chairman), H. Brown, J. Dodds (Secretary) T. Evans,
                     P. Field
1. *Apologies for absence* – none
2. *Minutes of last meeting* – taken as read
3. *Matters arising* – none
4. *Enquiries from prospective clients*
   4.1 *Westchester Hospital*
       Project: Air conditioning, new wing, Maternity
                 Ward block
       Approximate value; £100,000
       Tender date: 20th August
       Action: Secretary to write agreeing to tender.

   6.3 *Weekend Working, Roedale Road School*
       H.B. reported failure of Permalag to complete lagging and departure of that
       company's men from site and H.B.'s failure to contact their office. He requested
       help from P.F.'s men to work over the weekend and ensure completion on
       schedule.
       Action: P.F. confirmed that his men could cope and he was asked to arrange
       accordingly. H.B. to write to Permalag advising them of their failure and liability
       for costs.

Next Meeting: 09.00 hrs Friday 25th June, Conference Room
Distribution: P.A., J.A., H.B., J.D., T.E., P.F.

Implementing decisions taken

This is the third phase of the meeting's progress and in many respects it is the most important because it should produce *results*. Up to this point there has been communication of ideas, generation of proposals and decisions have been made but it is *implementation* of the decisions which is the main aim of most meetings. The chairman is *responsible* for implementation but it is frequently the secretary who has the *authority* to ensure that the persons required to take action are aware of what is expected of them and that they actually do it.

Item 3 on the standard agenda is 'matters arising' (from the minutes of the previous meeting) and a task of the group when considering this item would be to review the results of any action taken since the previous meeting.

Sub-committees and working parties

These provide means of implementing decisions of the main 'committee'. If, for example, the management group decided to investigate the possibility of installing an incentive payments scheme for site staff they might ask a few members of their group to investigate ways and means and report back to the management group. The sub-committee or working party (the terms are more or less synonomous) would normally have authority to co-opt to their sub-committee persons who could assist them with their task. Such sub-committees are usually 'ad hoc' – set up for a particular purpose – and they are disbanded when their task has been completed.

Advantages of meetings

Having commenced this chapter by referring to their disadvantages it is appropriate to conclude it by discussing their advantages. These are as follows.
(a)   for meetings of the information and/or instruction-giving type, the advantage of a meeting for the transmitter is that he is able to see and hear the response which his audience makes. This response is spontaneous and the transmitter can, then, if he wishes, adjust his message to match the response. Furthermore, the audience may find a personal, oral transmission more acceptable than the black-and-white message of a memorandum. The interchange of ideas which follows the initial transmission may more than justify the higher cost of a meeting than would be incurred with a memorandum.
(b)   for generating and interchanging ideas, formulating decisions, marketing and selling, social and staff participation purposes a meeting is almost invariably a suitable medium because it takes into account the human factor. Without an appreciation of the importance of that factor the manager is unlikely to get the best from his colleagues.

Reference

Chappell, D. (1980) Just a minute, *Building*, 26 September.

Chapter 10

Incentive payment schemes
and unit costing

'**incentive.** Tending to incite, n.Incitement (*to* action, *to* do, *to* doing),
provocation, motive'

'**motive.** Tending to initiate movement, . . . what induces a person to act, e.g.
desire, fear, circumstance' (COD)

Obtaining satisfactory productivity from his work-force is one of the sub-contractor's major tasks. To be *satisfactory* the work must be sufficient in *quantity* and it must also meet the *quality* requirements of the specification.

To produce, man needs motivation and/or incentive.

'The stick or the carrot' have always been regarded as the alternative motivators. In former times the 'stick' was unemployment, the workhouse and starvation (it still is in some countries) and the 'carrot' was any inducement which could be held in front of 'the donkey' to make him go.

When the stick becomes ineffective it is necessary to turn to the carrot. For man, the carrot takes various forms.

Incentives and motivators

Motivation was discussed in Chapter 2 by way of introduction to aims, objectives and corporate plans and the comments made in that chapter should be kept in mind. In the work environment several incentives may be identified.

1. *A congenial environment* where the man is working with his mates as part of a team and is therefore high on the list of motivators. Man likes to 'belong' and contribute.
2. *Competition* between groups of men working on similar tasks on the same site is frequently an incentive to increased productivity although it is often difficult to sustain the aura of competition over a prolonged period because one or the other of the teams tends to give up when it consistently fails to win.
3. *Security of employment* can be an effective motivator particularly in times of high unemployment.
4. *Promotion prospects* are not always effective as a motivator for operatives but for supervisors and managers the status and the symbols of status which accompany promotion are frequently greater incentives to improved performance than the salary increase which accompanies the promotion.
5. *Money* is the obvious motivator. Research has shown that it is by no means the most important motivator but it is, nevertheless, one which is used in a production context and it provides the main subject of this chapter.

Types of payment schemes not linked to productivity

1. *Basic rates* In the absence of an incentives payment based on improved levels of productivity operatives are generally paid on basic hourly rates of pay which have been agreed between the federations, associations and trade unions which negotiate between the employers and employees.
2. *'Plus' rates* These are a crude form of incentive payment. The operator is paid more than the basic rate as an inducement to (or reward for) increased productivity. Plus rates are frequently paid to operatives when the building industry has an excessive work-load as inducement to them to stay in their present employment. 'Plus' rates are payment for 'being there'.
3. *'Standing' bonus payments* are similar in their aims to plus rates but they are not linked to the hours worked by the operative. They are a lump sum paid, usually, weekly.
4. *Benefits* such as free medical care, free use of company facilities (sports, holiday flats, yachts, etc.) aim to hold the employee to the company.

Types of productivity-based incentive payments schemes

1. *Profit-sharing* schemes are based on project or corporate profits over a considerable period of time, e.g. half-year or year. They are usually regarded as more appropriate for management than work-force.
2. *Co-partnership* schemes are similar to profit-sharing schemes. The employees are technically 'partners' and they have a shareholding.
3. *Piece-work* payments are made to self-employed men. Piece-work is frequently known as 'the lump'. A lump sum is agreed in advance for a 'piece' of work to be carried out and it is paid on completion of the work.
4. *Earned bonuses* are similar to piece-work payments in that they are directly and immediately related to the operative's output but are paid *in addition* to the basic earned bonuses.

Earned bonus schemes

For any scheme to succeed there must be a willingness on the part of all concerned for it to be introduced. A scheme is not a substitute for efficient management. It should be integrated within the company's management system and it should be tailored to fit the functions of a particular company.

Reasons for introducing a scheme

The primary reason is usually to improve the men's earnings, to improve their efficiency and to encourage them to arrive at the place of work on time and work the full day.

Effects on management

The introduction of a scheme imposes a discipline on all the management staff involved. Managerial efficiency is improved because operatives' earnings are likely to suffer if the work has not been organized in such a way that they will be able to push ahead. In this event they will make their feelings known!

The sub-contractor must ensure that materials and plant are available as and when needed, that the site has reached a state of advancement so that the items to be installed *can* be installed without delay, that the materials are located to prevent unnecessary double-handling, that plant and a suitable power source are available, that enough but not more than enough men are sent to the site to do the job, that quality control arrangements are in hand, that men are not delayed by adverse weather conditions (there may be a choice of sites: one exposed, one with inside work) and that job B is awaiting the men when they have finished job A. For short-term jobs a Job Assignment Sheet of the type illustrated in Fig. 7.6 (p. 114) makes provision for calculating earned bonus by comparison of the *estimated job duration* and the (actual) *total time*.

Managerial advantages

Managerial advantages for accepting the disciplines imposed mean that recruitment becomes easier, labour costs are more easily controlled, output is improved and turnover is increased. Furthermore, the introduction of a scheme is often the beginning of an integrated management system which can have a significant effect on the efficiency of the company.

Managerial disadvantages

Managerial disadvantages are that the introduction of a scheme may lead to labour problems if management does not take care to provide the right conditions for the operatives to earn a bonus, and that there may be cost incurred initially in setting up the system to cope with the scheme. Closer supervision is necessary to ensure that quality of workmanship is not reduced because men have skimped work in order to earn more money.

Tasks involved in operating a scheme

Most schemes rely on providing the operatives with a *target cost* for carrying out a given task, performing a specified operation. The time taken by the operatives to perform the task is measured and the *actual cost* of performing the task is calculated. The operative is paid the difference between target cost and actual cost, assuming the former exceeds the latter. If the actual cost exceeds the target cost the operative is paid only his basic rate. Any deficit may be deducted from bonus monies owing to the operatives from retention money held by the sub-contractor or deducted from bonus earned in the future. In any event the operative has the security of his basic rate.

The tasks involved in operating a scheme comprise:

(a) setting targets (or budgets);
(b) measuring output;
(c) calculating the costs and the values;
(d) calculating the bonus or loss;
(e) negotiating with the operatives.

Task (a) requires judgement and knowledge of estimating or method study. Task (b) requires measuring skill. Tasks (c) and (d) involve some arithmetical skill and task (e) requires tact, integrity and negotiating skills.

Setting targets

This is the first task. For practical purposes budgets are usually prepared so that they may be used for both cost control and earned bonus target purposes. Experience shows that for both purposes weekly reviews are appropriate. Having decided on the review period the labour targets may be allocated to the time-scale in the manner shown in Fig. 7.7 (p. 119) and discussed in Chapter 7.

Measuring output

Measuring output for the period is the second task. Measurement may be made physically, on site or from drawings or it may be calculated as a percentage of the total work contained in the operation.

Calculating actual value

Calculating actual value or work carried out is the third task. This may be done from knowledge of the cost of each unit which makes up the whole of an operation or by calculating the value of the part of the operation carried out.

For example: referring to Fig. 7.7, assuming that the measurement is being carried out at the end of the week ending 26 April and that the measurements show that 40 per cent of the work to the riser had been completed. From the tender summary it is known that the *total* labour value of the completed riser is £3,125. The value of work completed to date is, then, 40 per cent of £3,125 £1,250. From this figure must be deducted the value of work previously completed, £890 in this instance, to give a labour value for week ending 26 April of £360.

Calculating the cost of the labour

Calculating the cost of the labour is the fourth task. In order to calculate the cost it must first be 'measured' and this is best accomplished by recording the time

		DAILY ALLOCATION SHEET for CONTRACT Excel Offices										Day Tuesday Date 16th April Week No.
		OPERATION	Switch room	Riser	Ground Floor Mains and Cables	Temporary wiring for Genr Contr.	Revision to Switch positions					
OPERATIVES		Opn. No.	28−220	30−210	38−47	Order no 15	VO 7					Total hrs
No.	Name	Grade	HRS. PER OPERATION									
	T Pearce	Tn	2	2	4							8
	G Meyer	E	8									8
	F Widdicombe	E	8									8
	W Brewer	E			4	4						8
	D Stewer	AE			4	4						8
	P Gurney	AE		8								8
	P Davey	AE				8						8
	D Widden	App				8						8
	H Hawk	App			4	4						8
	T Cobley	Lab			4	4						8
	Total	Hrs	18	10	20	16	16					80

Fig. 10.1 Recording and allocating labour to operations

spent by the work-force on the various operations. A Daily Allocation Sheet of the type shown in Fig. 10.1 may be used for this purpose. The pro forma is completed *daily* by the foreman, manager or chargehand on the site. The spaces for 'day' and 'date' act as a check on each other. Most chargehands know which day of the week it is but they are not always so confident about the date. If those entries tell conflicting stories it is usually safe to assume the *day* is correct and adjust the date the odd day or two to agree with the day.

The week no. may be inserted by the person responsible for the payment calculations.

Records of variations and dayworks

The 'operations' are those used in the programme or whatever operations the contract engineer (manager or proprietor) considers to be appropriate. What is most important is that once the operations have been determined *all* the operatives' time is allocated by the chargehand to one or other of the operations. Exceptions to this rule are items arising from variation orders or dayworks. This raises an important aspect of the use of daily allocation sheets. Because the chargehand *has* to allocate his operatives' time he *automatically* records time to 'variation' items and dayworks. Without this discipline the term spent on varied items might well be lost by the sub-contractor. The significance of this will be apparent to sub-contractors who have experience of contracts which have suffered from excessive variations. Further reference to this is made in Chapter 12.

In Fig. 10.1 the operations recorded are shown in the bar-chart referred to above (Fig. 7.7) plus a 'variation'.

The total hours spent by each operative, in the right-hand column, must be the sum of the times allocated to all the operations and the total must be the hours actually spent on site. This may appear to be an obvious statement but it is surprising how often errors occur.

At the end of each week the hours for each operation are abstracted by the proprietor or estimator or whoever in the firm is responsible for costing and/or estimating from the daily allocation sheet. The Method Statement for the operation, of the type shown in Fig. 4.2 (p. 45) may be used for the abstract. From the abstract of hours for the operation the actual labour cost may be calculated. Table 10.1 shows typical bonus calculations for the *ground floor mains and cables* operation. In practice the sub-contractor usually finds it necessary to use pro forma for recording the calculations.

If an operation can be completed in a single week it is a comparatively simple matter to compare budgeted with actual labour costs, but if, as is the case with the ground floor mains and cables, the operation extends over more than one week it is necessary to measure the quantity of work in order to make a true comparison of cost with value.

The word 'measure' may give the wrong impression because a calculation of the percentage of the operation which has been completed may be sufficiently accurate for valuation purposes. The percentage may be calculated from, say:

No. of units installed to date
        Total to be installed

Ideally, an operation should be capable of completion in one week or less but operations with a duration of more than a week can usually be broken down reasonably accurately.

Table 10.1 Typical bonus calculations

Ground-floor mains and cables
Bonus calculations

Week ending 12 April;		
Work value 35% × £855 (budget)		299.25
Actual cost (as abstract)		319.25
bonus/loss		(20.00)
		(no payment)
Week ending 19 April:		
Work value – as budget		855.00
Actual costs w/e 12 April	319.25	
(as abstracts) w/e 19 April	460.75	780.00
		75.00
Less previous bonus payment		Nil
bonus/loss		£75.00
		(operation complete)

Cost feedback for future estimating

The *actual cost* of an operation can provide useful feedback for future estimating. Care should be taken when considering feedback from an individual operation to ensure that it is not given too much significance but over a period of time a contractor's cost data from a representative sample of previous projects can be of value for the future.

The mechanics of incentive payment systems are not complicated but it will be appreciated that simplified examples have been used above. Care should be taken when setting up a system – quite a lot is at stake and it may be worth seeking outside advice.

The human aspects

The human aspects of installing an incentive payments scheme are extremely important and the sub-contractor ignores them at his peril.

Some points to have in mind are:
(a) start with a pilot scheme on a contract with a cooperative foreman. (If it succeeds on one project it should succeed on others);
(b) keep the scheme simple;

(c)   ensure that the operatives understand the reasons for installing the scheme and the scheme itself;

(d)   endeavour to install a 'fraud-proof' scheme – if it is sufficiently simple it will be less susceptable to fraud;

(e)   all transactions between management and operatives must be scrupulously fair and be seen to be so.

The introduction of an incentive payment scheme is often the forerunner of an integrated management system involving contract planning and budgetary control which may be of great benefit for the future of the firm.

Chapter 11

Architect's instructions and contractor's directions

Architect's instructions to contractor

The contractor's position with regard to architect's instructions is set out in clause 4.1.1 of SF 80 which reads:

4.1   .1   The Contractor shall forthwith comply with all instructions issued to him by the Architect in regard to any matter in respect of which the Architect is expressly empowered by the Conditions to issue instructions; save that where such instruction is one requiring a Variation within the meaning of clause 13.1.2 the Contractor need not comply to the extent that he makes reasonable objection in writing to the Architect to such compliance.

The clause is similar to clause 2(1) in the 1963 edition of the Standard Form of Building Contract in intention although the wording regarding compliance (or rather the need not to comply) in the penultimate line of SF 80, clause 4.1.1 is new.

This chapter concentrates on the 'instruction' rather than the 'variation' part because variations are the subject of the next.

Contractor obliged to comply with instructions

The contractor is required to '*comply with all instructions issued to him by the architect*' provided the architect is not exceeding his powers in issuing the instructions.

If the contractor wishes to be satisfied that the architect *is* empowered to issue the instruction in question, SF 80, clause 4.2 entitles him to request the architect to specify in writing the provision of the conditions of contract which empowers him to issue it. This entitlement occupies ten lines in the conditions of contract but it is seldom that a contractor finds it necessary to invoke it. In effect, the architect is empowered to issue instructions on just about anything and the main contractor is obliged to comply with them but this is not to say that the contractor will not be paid for compliance with an instruction! Quite the contrary – there is provision for payment as will be seen later.

If within seven days after receipt of a written notice from the architect requiring compliance with an instruction the contractor does not comply with it the *employer may employ and pay other persons* to execute any work which may be necessary to give effect to such instruction and all costs incurred in connection with such employment may be recovered from the contractor (SF 80, cl. 4.1.2).

Confirming oral 'instructions'

Perhaps of more importance to the nominated sub-contractor than the above 'safety-nets' is the procedure for dealing with 'instructions given otherwise than in writing'. This is mentioned because it may assist the nominated sub-contractor to more understanding of what sometimes appear to be inordinate delays in the issue of instructions. Furthermore, the sub-contractor has to follow very similar procedures. SF 80, clause 4.3.1 states that *'all instructions issued by the Architect shall be issued in writing'*.

So what happens if the architect 'purports to issue an instruction otherwise than in writing'? Such an instruction has no immediate effect but *shall be confirmed in writing by the contractor to the architect within seven days* and if not dissented from in writing by the architect to the contractor within seven days from receipt of the contractor's confirmation shall take effect as from the expiration as the latter said seven days.

An interesting feature of that condition, (which has been taken almost, but not exactly word for word from cl. 4.3.2), is the fact that the second *seven days period* runs from *'receipt* of the contractor's confirmation' (my italics). In the pre-second-class mail days, seventeen days was usually regarded as a reasonable allowance for the whole procedure (7 + 3 + 7 days), but with present day postal services ...!!

There is nothing to prevent the architect from giving an oral instruction on site or over the telephone and then confirming it in writing and in this event the 'instruction' takes effect from the date of the architect's written confirmation (SF 80, cl. 4.3.2.1). In this event, the main contractor is not obliged to confirm the instruction.

Architect's and contractor's oversight

There is a safeguard in case both contractor and architect fail to follow the procedures regarding confirmation of oral instructions discussed above but nevertheless the contractor goes ahead and, say, puts work in hand regardless. This lamentable oversight occasionally comes to light when either the contractor or the quantity surveyor is preparing the final account and discovers that there is no authority for including an item in the account. In this event the architect may, belatedly, confirm the instruction in writing 'at any time prior to the issue of the Final Certificate and the said instruction shall thereupon be deemed to have taken

165

effect on the date on which it was issued otherwise than in writing by the architect' (SF 80, cl. 4.3.2.2).

It is, however, an imprudent contractor who relies on clause 4.3.2.2 – it is far safer to ensure that oral 'instructions' are confirmed at the time. Much can happen between the time of issue of an instruction and the preparation of the final account and if the client's budgets have been exceeded an architect might be tempted *not* to give the confirmation desired by the contractor.

Architect's instructions to the nominated sub-contractor

Clause 4.2 of sub-contract NSC/4 reads:

4.2   The Contractor shall forthwith issue to the Sub-Contractor any written instruction of the Architect issued under the Main Contract affecting the Sub-Contract Works (including the ordering of any Variation therein); and may issue any reasonable direction in writing to the Sub-Contractor in regard to the Sub-Contract Works.

In Chapter 6 reference was made to the requirement contained in NSC/4, clause 4.1.2 for the sub-contractor continually to keep upon the sub-contract works a competetent person-in-charge. Any written instruction of the architect or direction of the contractor given to the person-in-charge is deemed to have been issued to the sub-contractor.

What is 'reasonable?'

There is always the question of what constitutes a 'reasonable' direction of the contractor, referred to in clause 4.2 above. Lawyers have trouble defining the meaning of 'reasonable'. Satisfying 'the average man on the Clapham bus' is frequently regarded as a measure of reasonableness. It is to be hoped, then, that the sub-contractor will not encounter unreasonable directions or he may have to rely on an arbitrator's interpretation and the settlement of disputes by arbitration (for which there is provision in art. 3 in NSC/4) can be a costly business although SF 80 does make provision for 'instant' arbitration. Previously the parties had to wait until the contract had been completed.

A point to note in NSC/4, clause 4.2 is that architect's instructions 'shall forthwith' be issued to the sub-contractor by the contractor. Once the contractor and sub-contractor have entered into contract communications between the architect and sub-contractors should all be via the contractor.

The sub-contractor's obligation to comply with any instruction or direction is set out in NSC/4, clause 4.3. It is similar to SF 80, clause 4.1.1 in much of its wording and the comments made in reference to SF 80, clause 4.1.1 apply here.

The procedures for dealing with instructions or directions 'otherwise than in writing' are contained in NSC/4, clause 4.4 and they are similar to those in SF 80 particularly with regard to the seven-day periods for confirmation.

If all the procedures are taken to the limit, the issue of instructions or directions can be a very time-consuming matter. All parties should use their common sense and best endeavours.

Sub-contractor at risk

It is often tempting for the nominated sub-contractor to ignore the conditions of contract and put work in hand before he has a direction from the contractor or an architect's instruction in writing. Indeed, it is often said that if all parties worked to the letter rather than the spirit of the contract they would never actually get the job done. This attitude of expediency is commendable and when contractors and sub-contractors have established a rapport with the architect they are usually safe in taking an architect's word as being as good as his written instruction. We are all 'honourable men'. Nevertheless, the sub-contractor *is at risk* if he fails to honour the letter of the contract, and obtain written confirmation. The architect may walk under a bus or the contractor may misinterpret his instruction. Most contractors keep a 'confirmation of oral instructions' book in which they write the architect's instruction and ensure that he signs it before he leaves the site. Sub-contractors should be able to work out a similar system.

Directions from contractor

The nominated sub-contractor is vulnerable if he fails to comply with a direction from the contractor. Clause 4.5 of NSC/4 reads:

4.5   If within 7 days after receipt of a written notice from the Contractor requiring compliance with a direction of the Contractor the Sub-Contractor does not begin to comply therewith, then the Contractor may, if so permitted by the Architect, employ and pay other persons to comply with such direction and all costs incurred in connection with such employment may be deducted from any monies due or to become due to the Sub-Contractor under the Sub-Contract or shall be recoverable from the Sub-Contractor as a debt.

Here, the sub-contractor is liable to incur costs in connection with employment of a person called in to comply with a direction with which the sub-contractor has not complied.

Clause 4.6 of sub-contract NSC/4 is the sub-contract counterpart of SF 80, clause 4.2. SF 80, clause 4.6 entitles the sub-contractor to request the architect to specify in writing the provision of the main contract which empowers the issue of an instruction. It also entitles the sub-contractor to use the contractor's name (and if necessary for the contractor to join with the sub-contractor), in arbitration proceedings at the instigation of the sub-contractor.

The procedures to be followed with regard to architects' instructions and contractor's directions are rather tedious and the sub-contractor should *read the conditions most carefully* when instructions and/or directions are in the offing.

Chapter 12

Variations and their valuation

By no means all the architect's instructions or contractor's directions which were discussed in Chapter 11 give rise to variations. What, then, *is* a variation?

Defining 'variation'

Unlike the old *green form*, norminated sub-contract NSC/4 contains in clause 1, a heading: 'Interpretation, definitions etc.' Many of the definitions etc., are not very helpful. Neither party to the contract may really be much wiser from reading that 'Arbitrator:[is] the person appointed under article 3 to be the Arbitrator', and on occasions they may be slightly confused to read that a 'person [is] an individual, firm [partnership] or body corporate'. (In both quotations, above, the [is] is my insertion).

When it comes to 'variation', however, NSC/4 is almost too effusive and we find:

where clause 15.1 applies the term 'Variation' means any of the following changes which are required by an instruction of the Architect:

.1　the alteration or modification of the design quality or quantity of the Sub-Contract Works as shown in the Sub-Contract Documents including:
  .1　the addition, omission or substitution of any work,
  .2　the alteration of the kind or standard of any of the materials or goods to be used in the Sub-Contract Works,
  .3　the removal from the site of any work, materials or goods executed or brought thereon by the Sub-Contractor for the purposes of the Sub-Contract Works other than work, materials or goods which are not in accordance with the Sub-Contract:

.2　the addition to, alteration of or omission of any obligations or restrictions imposed by the Employer as set out or referred to in the Tender, Schedule, 1, item 13 in regard to:
  .1　access to the site or use of any specific parts of the site;
  .2　limitations of working space;
  .3　limitations of working hours;
  .4　the execution or completion of the work in any specific order.

Where clause 15.2 applies the term 'Variation' has the same meaning but in line 4 and 5 of this definition delete 'design quality or quantity' insert 'design or quality'.

A variation involves 'change'

It can be seen from the above definition that the key element of a variation is 'change' and, indeed, in the North American construction industry the equivalent of a 'variation order' is a change order!

Nominated sub-contractors may like to bear in mind that the definition of 'variation' given above is almost identical with that contained in SF 80, clause 13.

Employer's obligations and restrictions

Items .1.1 to .1.3 in the definition are essentially the same as the conditions contained in the 1963 edition of the main contract but items .2.1 to .2.4 are new.

Whilst it has generally been understood that any 'obligations or restrictions imposed by the employer' would justify the contractor or a sub-contractor seeking reimbursement for any costs and/or expense which he might incur as a result of their imposition, the principle has not been stated as such before. Examples of obligations or restrictions which might be covered by items .2.1 to .2.4 are restriction of access to or through certain rooms during the alterations of, say, an occupied hotel or office building; limitation of working hours to evenings or weekends to suit the employer's convenience or a restriction imposed by the employer, which made it necessary for the sub-contractor to install his works in such a manner that certain parts of the building could be occupied and used before the sub-contract works were completed as a whole.

Lump sum and remeasurement sub-contracts

The definition of a variation refers to NSC/4, clauses 15.1 and 15.2 so it is obviously essential to refer to those sub-clauses:

15.1   Where in the Tender the Sub-Contractor has quoted a VAT-exclusive Sub-Contract Sum the price for the Sub-Contract Works shall be the Sub-Contract Sum or such other sum as shall become payable in accordance with the Sub-Contract.

15.2   Where in the Tender the Sub-Contractor has quoted a VAT-exclusive Tender Sum subject to complete re-measurement the price for the Sub-Contract Works shall be such sum or sums as shall become payable in accordance with the Sub-Contract and the total of such sum or sums shall be the Ascertained Final Sub-Contract Sum.

The essential difference between the sub-clauses is that 15.1 assumes a *lump sum* which would probably have been based on drawings and a specification and

169

15.2 suggests the existence of a *Schedule of rates* which the nominated sub-contractor has priced. They are, of course, *alternative* methods of submitting a tender. Assuming, then, that an architect's instruction has given rise to a variation, how will the effect of the variation be valued so that the sub-contractor may be reimbursed?

Valuation of variations

Clauses 16 and 17 in nominated sub-contract NSC/4 are concerned with the valuation of variations and as, between them, the clauses spread over more than two-and-a-half pages the sub-contractor will realize that valuation is not a matter to be taken lightly.

Variations to the sub-contract sum

Where in his tender the sub-contractor has quoted a sub-contract *sum* all variations and all work executed by the sub-contractor in accordance with instructions of the architect as to the expenditure of provisional sums included in the sub-contract documents shall be valued by the quantity surveyor in accordance with the provisions of clause 16.3.

Variations to a sub-contract *sum* will usually take the form of:

(a) additional or substituted work which *can* properly be valued by measurement; and/or

(b) additional or substituted work which *cannot* properly be valued by measurement; and/or

(c) omission of work which is set out in bills of quantities or other documents.

For the time being variations arising from instructions of the architect as to the expenditure of provisional sums may be ignored because they are a rather different matter from categories (a)–(c) above.

Valuation by measurement

The rules for the valuation of additional or substituted work which can properly be valued by measurement are given in clauses 16.2 and 16.3 of NSC/4 as:

16.2   Where the Sub-Contractor has included in the Tender a schedule of rates or prices for measured work and/or a schedule of daywork prices, such rates or prices shall be used in determining the Valuation in substitution for any rates or prices or Daywork Definitions which would otherwise be applicable under the relevant provisions of clause 16.3.

16.3   .1   To the extent that the Valuation relates to the execution of additional or substituted work which can properly be valued by measurement such work shall be measured and shall be valued in accordance with the following rules:

.1　.1　Where the work is of similar character to, is executed under similar conditions as, and does not significantly change the quantity of, work set out in bills of quantities and/or other documents comprised in the Sub-Contract Documents the rates and prices for the work so set out shall determine the valuation.

.1　.2　Where the work is of similar character to work set out in bills of quantities and/or other documents comprised in the Sub-Contract Documents but is not executed under similar conditions thereto and/or significantly changes the quantity thereof, the rates and prices for the work so set out shall be the basis for determining the valuation and the valuation shall include a fair allowance for such difference in conditions and/or quantity.

.1　.3　Where the work is not of similar character to work set out in bills of quantities and/or other documents comprised in the Sub-Contract Documents the work shall be valued at fair rates and prices.

Clause 16.2 makes it quite clear that where there is a schedule of rates or prices for measured work, etc., such rates or prices shall be used in determining the valuation.

Clause 16.3.1 requires there to be 'bills of quantities' or 'rates and prices' for the work. In this event the quantity surveyor's task is reasonably straightforward. If it is necessary for him to *measure work for the purpose of the valuation*' clause 16.4 states that '*the Contractor shall give the Sub-Contractor an opportunity of being present at the time of such measurement and of taking such notes and measurements as the sub-contractor may require*'.

It is a nice point if measurement of 'work' includes measurements which the quantity surveyor may take from *drawings* which show the varied work.

Whatever the legal interpretation may be, most quantity surveyors would no doubt welcome the sub-contractor's engineer being present at the time of measurement regardless of how the measurements were taken. Having measured the work to be valued the quantity surveyor is able to use the rates and prices in the bills of quantities or other documents to determine the valuation in accordance with the first and second rules set out in clauses 16.3.1.1 and 16.3.1.2.

The first rules applies where the work is '. . . of similar character to, is executed under similar conditions as, and does not significantly change the quantity of . . . ' and in this event the rates and prices for the work (in the bills of quantities, etc.) '*determine* the valuation'. The second rule applies where the work *is not executed under similar conditions to, or is significantly different in quantity from,* the work which the sub-contractor tendered to carry out. In this event the rates and prices etc., provide '. . . *the basis* for determining the valuation and the valuation shall include a fair allowance for such difference in conditions and/or quantity'.

Reference should be made to the use of the word 'similar' when it is used in the context of clauses 16 and 17.

The *Oxford English Dictionary* says of similar: 'of the same substance or structure throughout' but in recent discussion regarding SF 80 it has been argued that similar means 'identical'. The argument goes that a contractor contracts to carry out work *identical* to the work in the bills of quantities *not closely resembling it* or *similar* to it. The counter argument is that the intention of the contract is that the work arising from the variation need not be *the same in every detail* in order that its value may be determined from the rates or prices for the work contained in the bills of quantities etc. This is a rather academic argument on a par with *'what would satisfy a reasonable man'* but there appears to be more to be said for the 'close resemblance' than the 'identical' school. The need for precision in legal drafting is obviously essential but valuation is not an exact science and it is valuation that is discussed in this instance.

Reference was made above to using the rates and prices for the work in the bills of quantities, etc. to *determine* the value of variations where the work is of similar character to, is executed under similar conditions as, and does not significantly change the quantity of, work set out in the bills of quantities or other sub-contract documents (NSC/4, cl. 16.3.1.1). Where, however, the work is of similar character to work set out in bills of quantities or other sub-contract documents *but is not executed under similar conditions*, etc. (see NSC/4, cl. 16.3.1.2 for the full details) the rates and prices for the work so set out shall be *the basis for determining* the valuation and the valuation shall include a fair allowance for such difference in conditions and/or quantity.

The quantity surveyor's task

What does all this mean in practical terms?

The quantity surveyor's task is to prepare a final account for all the work carried out by the sub-contractor. This account takes in any variations arising from architect's instructions. He, the quantity surveyor, will be exceeding his authority if he includes in the account items which *do not* arise from architect's instructions. It is, therefore, common sense and good practice for the sub-contractor to ensure that written architect's instructions are in existence.

The sub-contractor should, also, be aware of the format that the quantity surveyor will use when preparing his final account so that he, the sub-contractor, may provide the quantity surveyor with information in a form most suitable for inclusion in the final account. In Chapter 15, Table 15.1 (p. 232), there is an example of a simple final account format. *It is up to the sub-contractor to check and agree or disagree with the quantity surveyor's final account.* The previous sentence has been underlined to emphasize its importance.

Refer to item 2 in the final account format in Table 15.1 which shows, in 'variation', the way in which the quantity surveyor calculates the sums which appear in the omissions and additions columns.

Omissions

Sub-contractors are occasionally ill-informed about omissions.

They are referred to in NSC/4, clause 16.3.2 which says:

16.3 .2 To the extent that the Valuation relates to the omission of work set out in bills of quantities and/or other documents comprised in the Sub-Contract Documents the rates and prices for such work therein set out shall determine the valuation of the work omitted.

Omissions comprise items of work which were included in the sub-contractor's tender but which it has been decided will not be carried out. An example would be an employer's decision not to have light fittings installed as part of the contract works but to leave the tenants to make their own arrangements in that respect.

It is unlikely that the sub-contractor will have much to disagree with in the omissions but it is not unknown for the quantity surveyor to omit work which has, in fact, been carried out. This may be because an item was omitted at one stage of the contract (which the quantity surveyor knew about) but subsequently an oral instruction was given to go ahead with it after all, the oral instruction not having been confirmed.

Valuing variations from bills of quantities

Omissions are, however, only part of the varied work which has to be valued. The method of valuing varied work which is described in NSC/4 clauses 16.3.1 and 16.3.2 is set out in clause 16.3.3 which reads:

16.3 .3 In any valuation of work under clauses 16.3.1 and 16.3.2:

.3 .1 where bills of quantities are a Sub-Contract Document measurement shall be in accordance with the same principles as those governing the preparation of those bills of quantities as referred to in clause 18.

.3 .2 allowance shall be made for any percentage or lump sum adjustments in bills of quantities and/or other documents comprised in the Sub-Contract Documents; and

.3 .3 allowance, where appropriate, shall be made for any addition to or reduction of preliminary items of the type referred to in the Standard Method of Measurement 6th Edition, Section B (Preliminaries).

Clause 16.3.3, it will be noted, applies where bills of quantities are a sub-contract document.

When the quantity surveyor has a sub-contractor's bills of quantities as a means of valuing varied work his task is, as stated above, reasonably straightforward, particularly regarding the valuation of work which has been omitted.

When it comes to additions his task is rather more difficult. This is where he has to value along the lines referred to in NSC/4, clause 16.3.1 and the sub-contractor may wish to ensure that the quantity surveyor has sufficient information to enable him to value items of work for which there are not identical rates in the bills of quantities.

Generally speaking, quantity surveyors are more knowledgeable about rates for building works than rates for some items of specialist contractor's work and they will often be receptive of assistance from the sub-contractor.

Pro rata rates

The time-honoured method used by quantity surveyors for pricing items of work which are similar to, but not identical with, items in the bills of quantities is the pro rata basis.

As an elementary example of this method assume that the bills of quantities contain two items for, say, cable trunking which the sub-contractor had priced as:

100 mm × 50 mm cable trunking	290 m	£3.50	£1,015.00
100 mm × 100 mm trunking	400 m	£5.50	£2,200.00

and the architect has given an instruction which gave rise to a variation for 50 m of 100 mm × 75 mm trunking. Pro rata, the quantity surveyor might well price

the 100 mm × 75 mm trunking at £4.50 $\dfrac{(£3.50 + £5.50)}{2}$ and this calculation

may be fair and reasonable in most instances.

But quantity surveyors are not omniscient, (as much as it grieves me to admit it), and the sub-contractor may be able to provide information which would lead (or would have led) the quantity surveyor to value the 100 mm × 75 mm trunking differently. The sub-contractor may, for instance, have to pay a higher rate for trunking when purchasing less than a certain quantity – there is only 50 m of the varied trunking in the example – or the trunking may have been fixed to a surface which made it more costly (or for that matter less costly) for the sub-contractor. These are factors which a quantity surveyor will normally take into account if the sub-contractor brings them to his attention.

The constituents of items of the sub-contractor's work are labour, plant and materials from which it follows that the rates and prices which make up his tender should reflect the costs of those constituents. The quantity surveyor will not know the price build-up of the sub-contractor's rates unless the sub-contractor informs him of them.

In the above example, where the varied work comprises cable trunking of a size which slots comfortably between two other sizes, the quantity surveyor has good data for his valuation. If, however, the sub-contractor's tender had included only, say, the 100 mm × 50 mm cable trunking, £4.50/m for the 100 mm × 75 mm trunking may be an incorrect pro rata rate.

If it is assumed that the price build-up for the 100 mm × 50 mm cable trunking is:

	(£)
material	3.00
labour	0.50
	£3.50/m

and the material cost of the 100 mm × 75 mm cable trunking is £4.50/m the rate for the larger trunking would be:

	(£)	
material	4.50	
labour	0.50	(assuming it cost no more to fix the larger trunking)
	£5.00/m	

i.e. more than the pro rata rate of £4.50/m.

In fact, the difference in material cost between the two sizes would probably be much less and in this event the quantity surveyor might have over-valued the variation.

If the sub-contractor received a final account which included an over-valuation he would be at liberty to inform the quantity surveyor of his generosity but as the contract is quite clear that it is the responsibility of the quantity surveyor to 'value', the sub-contractor need not be too reluctant to accept.

Obtain orders when carrying out temporary works for the contractor

Work which the sub-contractor may have done for the contractor by way of wiring or plumbing for site offices, storage buildings, canteens, etc., wiring for contractor's plant, such as cranes and hoists, or for temporary wiring in the building under construction will not be included in the quantity surveyor's final account. Contractors are often dependent on specialist contractors for advice and on their services as specialist contractors and they may find that their services in this connection are a useful source of income. They should, however, ensure that they have written orders from the contractor for all the work which he wishes them to undertake. The sub-contractor will be prudent if he undertakes such work only on an 'accepted estimate' basis and only if both parties are clear in their own minds that these items of work are quite separate from the work which is the subject of the sub-contract under nominated sub-contract NSC/4.

The two previous paragraphs are a digression from the main theme of this chapter, but they are items of considerable importance from the sub-contractor's point of view.

So far, this chapter has been concerned mainly with the valuation of work referred to in NSC/4, clause 16.3.1.1 as work of 'similar character', etc. What about the work referred to in clause 16.3.1.2 – that work which is 'not executed under similar conditions' to that set out in bills of quantities, etc. or which is significantly different in quantity?

This, it will be remembered from clause 16.3.1.2, is where the rates and prices '. . . shall be *the basis* for determining the valuation and the valuation shall include a fair allowance for such difference in conditions and/or quantity' (my italics).

The quantity surveyor is on less firm ground when it comes to valuation on 'the basis'. To some extent but by no means entirely the 'trunking' example given above is in this category.

The sub-contractor has more scope for negotiation of rates because what, as was questioned above, determines whether or not work has been carried out 'under similar conditions' to work set out in bills of quantities? Obviously, wiring for a socket outlet which the architect instructed should be installed after the main contractor's work in a particular part of the new building had been completed and which involved the sub-contractor in cutting into finished work in order that he could comply with the architect's instruction would not be 'under similar conditions'. The sub-contractor would be entitled to expect a substantially increased rate by way of the 'fair allowance for such difference' which the quantity surveyor is required to make in accordance with NSC/4, clause 16.3.1.2.

The sub-contractor would be well advised to keep a record of the costs which he actually incurred so that he could bring any shortfall in the quantity surveyor's valuation to his attention.

Daily allocation sheets

Daywork sheets are one method of recording such items of work but if the sub-contractor's foreman allocates his men's (persons'?) time on a daily allocation sheet such as that shown in Fig. 10.1 (p. 160) there is no need for a daywork sheet for an item of work of this sort. Indeed, daily allocation sheets are in many respects more credible than daywork sheets because the quantity surveyor is able to see how the *whole* of a man's time has been spent on the day in question. Even the most sceptical quantity surveyor would find it difficult to disregard an allocation sheet which was one of a regular series and which bore all the hallmarks, (appropriate grubbiness, consistent handwriting and literary style, etc.) of a contemporary record. The mechanics of preparing daily allocation sheets is discussed in Chapter 10.

There remains for consideration the work referred to in NSC/4, clause 16.3.1.3 which is 'not of similar character to work set out in bills of quantities and/or other documents comprised in the sub-contract documents (which) . . . shall be valued at fair rates and prices'.

Here again, daily allocation sheets can be useful in determining 'fair rates and prices'.

Opinions differ about the tactics of quantity surveyor/sub-contractor liaison. Some sub-contractors maintain that it is better to let the quantity surveyor make his valuation and present the sub-contractor with a final account without the sub-contractor becoming involved. It is argued that the quantity surveyor sometimes makes higher valuations than the sub-contractor would anticipate and, indeed, higher than the costs which the sub-contractor incurred. This is probably one of the occasions where the sub-contractor's knowledge of the quantity surveyor's proclivities is particularly useful.

Dayworks

Dayworks may be used where work cannot properly be valued by measurement. NSC 4, clause 16.3.4 describes the methods to be used for calculating the cost of dayworks and for recording the time spent upon the work:

16.3 .4 To the extent that the Valuation relates to the execution of work which cannot properly be valued by measurement the Valuation shall comprise the prime cost of such work calculated in accordance with the Definition or Definitions of Prime Cost of Daywork identified on page 1 of the Tender together with percentage additions to each section of the prime cost at the rates set out by the Sub-Contractor in the Tender.

Provided that in any case vouchers specifying the time daily spent upon the work, the workmen's names, the plant and the materials employed shall be delivered for verification to the Contractor for transmission to the Architect or his authorized representative not later than the end of the week following that in which the work has been executed.

The 'definition' of prime cost of daywork referred to in the above extract is that which has been agreed between the RICS and the NFBTE, ECA or HCVA, etc. as the case may be. The 'percentage additions' to cover the cost of the sub-contractor's overheads and profit on daywork items are inserted on page 1 of the tender and agreement form, NSC/1, by the sub-contractor when he is negotiating with the architect and contractor at the time of his tender (see Ch. 5 and Fig. 5.3).

Sub-contractors should take care to follow closely the procedure set out in the second paragraph of NSC/4, clause 16.3.4 regarding the submission of 'vouchers' (or daywork sheets, as they are more usually known) to the contractor.

'The end of the week following that in which the work has been executed', the time by which the vouchers must be in the hands of the architect or his authorized representative, appears at first sight to be adequate but in practice the sub-contractor may find it quite difficult to comply with the requirement. It should be noted that whilst the sub-contractor is required to enter the workmen's names, the plant and the materials spent daily on the daywork sheets, he is *not* required to enter prices for those items at the time the vouchers are delivered to the contractor for verification. It is advisable to make reference to the appropriate architect's instruction on the daywork sheet if only to assist the quantity surveyor with identification.

The architect's signature on daywork sheets is usually considered to provide the sheets with an aura of authority but responsibility for determining the method of valuation rests, ultimately, with the quantity surveyor. This is not to suggest that the quantity surveyor would disregard the requirement of NSC/4, clause 16.3.4, set out above, it is simply that it is he who decides which of the valuation rules given in clause 16.3.1 will be applied. Nevertheless, the record which a daywork sheet provides is useful even if it does not appear in the final account in its original form.

Variation causing 'substantial changes'

The nature of a 'variation' may be that in addition to varying the sub-contractor's work it substantially changes the conditions under which other work is executed. Under such circumstances the sub-contractor may incur additional costs in carrying out the *other work* which he could not have foreseen at the time he submitted his tender. Clause 16.3.5, given in full below, envisages this contingency and it will be seen that in such an event the *other* work shall itself be treated as a variation.

16.3 .5 If compliance with the instruction requiring a Variation or the instruction as to the expenditure of a provisional sum to which the Valuation relates in whole or in part substantially changes the conditions under which any other work is executed, then such other work shall be treated as if it had been the subject of an instruction of the Architect requiring a Variation under clause 13.2 of the Main Contract Conditions which shall be valued in accordance with the provisions of clause 16.

Sub-contractors should take a special note of this clause. It does not have an equivalent in the green form and it may provide the sub-contractor with a means of obtaining reimbursement for costs which in the past he may have found it necessary to write off as 'just one of those things'.

A fair valuation shall be made

Sub-clause 16.3.6, given in full below, appears to be a 'sweeping up' provision.

16.3 .6 To the extent that the Valuation does not relate to the execution of additional or substituted work or the omission of work or to the extent that the valuation of any work or liabilities directly associated with a Variation cannot reasonably be effected in the Valuation by the application of clauses 16.3.1 to .5 a fair valuation thereof shall be made.

Provided that no allowance shall be made under clause 16.3 for any effect upon the regular progress of the Sub-Contract Works or for any other direct loss and/or expense for which the Sub-Contractor would be reimbursed by payment under any other provision in the Sub-Contract.

Valuation of 'complete remeasurement' contracts

Variations relating to 'complete remeasurement' contracts are referred to in NSC/4, clause 15.2.

The whole of NSC/4, clause 17 is devoted to complete remeasurement variations under the heading 'Valuation of all work comprising the Sub-Contract Works' by quantity surveyor. The last three words, are not in parentheses because they appear in the margin note rather than the heading.

Most of the content of clause 17 is very similar in its effect to that contained in clause 16 (valuation of variations on 'lump-sum' contracts). The contractor is required to give the sub-contractor an opportunity of being present at the time when the quantity surveyor is measuring work and of taking such notes and measurements as he, the sub-contractor, may require (NSC/4, cl. 17.2) just as he is required so to do in NSC/4, clause 16.4.

Where the sub-contractor has included in his tender a schedule of rates or prices and/or a schedule of daywork prices, such rates or prices shall be used in determining the valuation of the variations 'in substitution for any rates or prices or Daywork Definitions which would otherwise be applicable' (NSC/4, cl. 17.3).

Rules for valuation

The 'Rules for Valuation', to use the wording in the margin note opposite clause 17.4, are similar to those contained in clause 16.3 but, strangely, the margin note opposite clause 16.3 reads 'Valuation Rules'.

To return to the rules for valuation. Where work is of similar character and it is executed under similar conditions as, and it does not significantly change the quantity of, work set out in bills of quantities, etc. *the rates and prices* for the work so set out *shall determine the valuation* (NSC/4, cl. 17.4.1).

It is appropriate to mention at this point that clause 18 requires that the bills of quantities referred to in clauses 16 and 17 have to be prepared in accordance with SMM and that if there is any departure from the method of preparation referred to above or any error in description or in quantity or omission of items then such departure or error shall be treated as if it were a variation. Any valuation of a variation shall be in accordance with the same principles as those governing the preparation of the bills of quantities.

Where work is of similar character but it is *not* executed under similar conditions and/or the quantity of work is significantly changed *the rates and prices* for the work so set out *shall be the basis for determining the valuation* and the valuation shall include fair allowance for such difference in conditions and/or quantity (NSC/4, cl. 17.4.2).

Where the work is not of similar character to work set out in bills of quantities *the work shall be valued at fair rates and prices* (NSC/4, cl. 17.4.3).

Adjusting preliminaries

Specialist contractors using earlier editions of the JCT *Guide* should refer to the corrigenda inside the front cover for the next item because the original NSC/4, clause 17.4.2.3 has been deleted and the following words have been substituted: 'Any amounts priced in the Preliminaries Section of the Sub-Contract Documents adjusted, where appropriate, to take into account any instructions of the Architect requiring a Variation or in regard to the expenditure of a provisional sum included in the Sub-Contract Documents shall be included . . .' in any valuation of work under clause 17.

For the benefit of sub-contractors who are not familiar with bills of quantities, the 'preliminaries bill' is the bill which includes items concerned with running the job rather than the work itself. Typically, the preliminaries bill makes it possible for the sub-contractor to put a price to items for his site engineer (the person-in-charge), storage buildings, plant, insurances and a host of similar items.

Clause 17 is, then, important to the sub-contractor because it *entitles him to an adjustment* of his site management and 'overhead' costs as part and parcel of being paid for a variation.

Alternative calculations for preliminaries

There are, at least, two schools of thought about how the sub-contractor should price preliminaries items in bills of quantities, having in mind the prospect of them being adjusted during the life of the project.

The 'price preliminaries in with the rates for the work' school argues that the cost of preliminaries item should be *calculated as a total sum* for the whole of the sub-contract works and *included as a percentage on each of the rates*. For example: if the sub-contractor calculates the cost of the labour and materials in the 'rates and prices' to be £50,000 and the cost of preliminaries is calculated to be £5,000 (splendidly convenient figures) then:

$$\frac{£5,000}{£50,000} \times 100 \quad = 10 \text{ per cent}$$

should be added to each of the rates to cover the cost of the running of the job. The sub-contractor's profit margin may be treated in a similar manner.

The argument goes that any variation which gives rise to an increase in the amount of work which the sub-contractor has to carry out will automatically include the cost of preliminaries because every item of work priced in the bills of quantities has a 'preliminaries' percentage added to it. Furthermore, if the sub-contractor had anticipated and allowed in his tender for, say, a twelve-week contract period and he completed the works in six weeks then there can be no question of his preliminaries being reduced.

The 'price preliminaries separately' school's argument is, of course, the counter to that expressed above.

This school's argument is that part of the work may be *omitted* by the architect as a variation and that in this event the sub-contractor loses not only his profit on the work but his preliminaries costs as well. Furthermore, if the sub-contractor is required due to site conditions outside of his control, to carry out his work over a longer period of time than he had anticipated in his tender, the cost of his preliminaries items is readily seen if it is kept separate and a proper adjustment may be made. The argument continues that many more projects are extended than are completed ahead of programme and that in the event of the sub-contractor carrying out his work in less time than he planned it is an unimaginative contract engineer who cannot find some reason to look in on the site after his men have to all intents and purposes completed their work in order to justify having included a longer contract period in his tender.

It is really a case of 'you pays your money and you takes you choice' but generally speaking the second school appears to have the stronger case particularly now that SF 80 openly recognizes the need to adjust preliminaries items.

Provision for dayworks

There is provision for variations to be 'valued' on a daywork basis, in accordance with the appropriate definition of prime cost of daywork as 'identified on page 1 of the Tender together with percentage additions . . .' provided that vouchers specifying the time daily spent upon the work, etc. are delivered for verification to the contractor not later than the end of the week following that in which the work has been executed (NSC/4, cl. 17.4.3). All this is similar to the provisions of NSC/4, clause 16.3.4.

Where an architect's instruction requires a variation or any instruction as to the expenditure of a provisional sum the rules for valuation are essentially the same as those contained in NSC/4, clauses 17.4.1.2 and 17.4.1–17.4.4 and 17.4.4 and 17.4.5.

Clause 17 concludes by making it clear that clause 17.4 does not apply to items of direct loss and/or expense due to disturbance of the regular progress of the works for which the sub-contractor would be reimbursed by payment under any other provision in the sub-contract (NSC/4, cl. 17.4.5).

The effect of variations generally

Some reference should be made to the role of variations by way of conclusion of this chapter. It is often suggested that it is too easy for the client (or the architect) to change his mind – vary the works whilst they are in progress – and that his doing so is a major factor causing delays and disruption of the works which are so prevalent on British building projects. It is pointed out that in countries where bills of quantities are not used and there is no provision in the contract conditions for the contractor's and sub-contractors' rates and prices in the bills of quantities to be used when valuing variations, the architect and client are obliged to think through the design *before* work commences on site because if variations occur the contractors can (and do), charge what they like for the variation.

Various methods of discouraging variations have been suggested but the most effective method is no doubt the provision of complete documentation before the parties enter into contracts coupled with a determination on the part of all concerned to carry out and complete the works as expeditiously as possible.

Chapter 13

Delays and extensions of time

It is a sad fact of building life that far too few building projects are completed on time. There are numerous valid reasons for this, not least adverse weather conditions.

Delays and extensions of time appear to be so generally accepted that the Standard Form of Building Contract devotes several pages to the procedures to be adopted when they occur. There is even a collective name for them, in SF 80 'Relevant Events'. It is important that sub-contractors are familiar with the procedures which must be followed if they are to ensure their contractual invulnerability.

Specialist sub-contractors who are familiar with the Green Form and the 1963 edition of the Standard Form of Building Contract do not have a great advantage over those less knowledgeable when considering extension of time – SF 80 and NSC/4 (the latter in particular) are different from earlier editions.

It is, however, necessary to consider the main contract in the first place because although the extension of time clauses are similar in both main and sub-contracts the lead-in to the subject only appears in the main contract.

The broad picture is that the contractor takes possession of the site on a given date and he undertakes to complete the works on or before the 'Completion Date' (SF 80, cl. 23.1). The actual completion date is agreed when the employer and contractor enter into their contract and the agreed date is written into the appendix to SF 80.

If the contractor fails to complete the works by the completion date the architect issues a certificate to that effect (SF 80, cl. 24.1).

Issue of the certificate referred to above makes the contractor liable to pay the employer liquidated damages.

Liquidated damages

An explanation of 'Liquidated Damages'. Liquidated damages are intended as a means of compensating the employer for any loss which he may incur as a result of the contractor failing to complete on or before the Completion Date. In the appendix to SF 80 there is an item 'liquidated and ascertained Damages . . . at

182

the rate of £......... per..........'. When the employer and contractor enter into their contract they fill in the blanks. The sum inserted must be reasonable. It must represent the damages which the employer will actually suffer by way of loss of rental income, increased costs of financing the project, etc. as a result of the contractor's failure. English courts have never held that liquidated damages should be treated as a means of *penalizing* the contractor. The blank spaces in the appendix are usually completed as £x per day, week or month as the parties consider appropriate.

Liquidated damages often involve large sums so contractors are understandably concerned that the architect does not issue a certificate to the effect that he, the contractor, has failed to complete the works by the completion date.

Relevant events

There is provision in SF 80 for the architect to make an extension of time if the progress of the works is delayed as a result of a number of 'Relevant Events' (SF 80, cl. 25.1). Twelve relevant events are listed in SF 80, clause 25.4. They include exceptionally adverse weather conditions, the contractor not having received in due time necessary instructions and delay on the part of nominated sub-contractors or nominated suppliers. NSC/4, the Nominated sub-contractor form, has a similar list of relevant events in clause 11.2.5 which will be considered in detail later.

Written notice from contractor

Clause 25.2.1.1 of SF 80 requires that if it becomes reasonably apparent that the progress of the works is being or is likely to be delayed the contractor shall give written notice to the architect of the 'material circumstances including the cause or causes of the delay and identify in such notice any event which in his opinion is a Relevant Event'. Where the material circumstances in the written notice include reference to a nominated sub-contractor the contractor is required to send a copy of the notice to the nominated sub-contractor (SF 80, cl. 25.2.1.2).

The contractor's written notice should include particulars of the expected effects of the relevant event to which he refers in his notice and he should also estimate the extent of the expected delay in the completion of the works beyond the completion date (SF 80, cl. 25.2.2.2). If it is not practicable for the contractor to give the above particulars at the time he gives his written notice he is required to do so as soon as possible after the notice and he is also required to send copies of the particulars to nominated sub-contractors.

Particulars kept up-to-date

Clause 25.2.2.3 provides for the particulars and estimate to be kept up-to-date and the contractor is required to give to the architect further written notices as may reasonably be necessary or as the architect may reasonably require. Here, again, copies must be sent to nominated sub-contractors.

Written extension of time

Having received from the contractor any notices and particulars and estimates the architect is obliged to decide if, *in his opinion*, any of the events stated by the contractor to be the cause of the delay is a relevant event and if the completion of the works is likely to be delayed by it. If so, the architect is obliged to give to the contractor, in writing, an extension of time by fixing such later date as the completion date as he, the architect, estimates to be fair and reasonable (SF 80, cl. 25.3.1). In his letter to the contractor the architect is obliged to state which of the relevant events he has taken into account and the extent to which he has had regard to any instruction which he may have given which required as a variation the omission of any work issued since the fixing of the previous completion date.

Action within twelve weeks

The architect is required to take the action described above within twelve weeks from receipt of the notice, particulars, etc. from the contractor or, if the completion date is less than twelve weeks away, the architect must act before the completion date (SF 80, cl. 25.3.1)

New points

Two new points, in particular, emerge from the foregoing requirements. The first is that the architect has the power to try to reduce the effect of delays which have arisen by omitting part of the works. The second is that he is required to take action (to fix a new completion date) within a stated period of time. In the 1963 edition of the Standard Form of Building Contract he is merely required to estimate the length of the delay 'so soon as he is able'. The action expected of the architect in clause 25.3.1 is, incidentally, referred to as the 'first exercise of his duty' so presumably the following action, set out in clause 25.3.2, may be regarded as his second exercise.

Earlier completion date

This clause enables the architect to fix an *earlier* completion date than that previously fixed if in his opinion the fixing of such earlier completion date is fair and reasonable having regard to any instructions he may have given requiring as a variation the omission of any work. The instruction which gives rise to the omission must have been given *after* the last occasion on which the architect made an extension of time and (cl. 25.3.6) no decision by the architect under clause 25.3 shall fix a completion date *earlier* than the date for completion stated in the appendix to SF 80.

Sweeping-up

Not later than the expiry of twelve weeks from the date of Practical Completion the architect has three courses of action which he may take as a sweeping-up exercise. These are:

(a) he may fix a completion date *later* than that previously fixed if in his opinion this would be fair and reasonable (SF 80, cl. 25.3.3.1);

(b) he may fix a completion date earlier than that previously fixed if in his opinion . . . etc. (SF 80, cl. 25.3.3.2);

(c) he may confirm to the contractor the completion date previously fixed, (SF 80, cl. 25.3.3.3).

He must not just let matters rest – he must either do (a), (b) or (c).

From the nominated sub-contractor's standpoint it should be noted that 'the Architect shall notify in writing to every Nominated Sub-Contractor each decision of the Architect under Clause 25.3 fixing a Completion Date'. SF 80 puts quite an administrative load on the architect and perhaps it is not surprising that architects generally have not greeted it with enthusiasm. But sub-contractors should be better informed of what is going on as a result of all the architect's exercises.

Sub-contractor's obligations

The sub-contractor's obligation to carry out and complete the sub-contract works and extension of time are set out in clause 11 of NSC/4.

The (nominated) sub-contractor undertakes to 'carry out and complete the Sub-Contract Works in accordance with the agreed programme details in the Tender, Schedule 2, item 1C, and reasonably in accordance with the progress of the Main Contract Works but subject to receipt of the notice to commence work on site as detailed in the Tender, Schedule 2, item 1C, and to the operation of Clause 11.2', (NSC/4, cl. 11.1). Clause 11.2 has the heading 'Extension of Sub-Contract time'.

Sub-contractor's written notice

NSC/4, clause 11.2 describes the action which should be taken if the sub-contract works are delayed. The first phase of the saga reads:

11.2 .1 .1 If and whenever it becomes reasonably apparent that the commencement, progress or completion of the Sub-Contract Works or any part thereof is being or is likely to be delayed, the Sub-Contractor shall forthwith give written notice to the Contractor of the material circumstances including the cause or causes of the delay and identify in such notice any matter which in his opinion comes within clause 11.2.2.1. The Contractor shall forthwith inform the Architect of any written notice by the Sub-Contractor and submit to the Architect any written representations made to him by the Sub-Contractor as to such cause as aforesaid.

All three parties are involved; the sub-contractor gives *written notice* to the contractor who is obliged to *inform* the architect and pass on the sub-contractor's 'representations'.

If it is practicable for the sub-contractor to provide all the 'particulars' which the contractor and architect need to know in order to exercise their respective duties he should provide these particulars with his written notice. If he is not able to do so at that time he should do so 'as soon as possible after such notice'.

Sub-contractor's particulars and estimate

The clause reads:

11.2 .1 .2 In respect of each and every matter which comes within clause 11.2.2.1, and identified in the notice given in accordance with clause 11.2.1.1, the Sub-Contractor shall, if practicable in such notice, or otherwise in writing as soon as possible after such notice:

  .2 .1 give particulars of the expected effects thereof; and

  .2 .2 estimate the extent, if any, of the expected delay in the completion of the Sub-Contract Works or any part thereof beyond the expiry of the period or periods stated in the Tender, Schedule 2, item 1C, or beyond the expiry of any extended period or periods previously fixed under clause 11 which results therefrom whether or not concurrently with delay resulting from any other matter which comes within clause 11.2.2.1; and

  .2 .3 the Sub-Contractor shall give such further written notices to the Contractor as may be reasonably necessary or as the Contractor may reasonably require for keeping up-to-date the particulars and estimate referred to in clause 11.2.1.2.1 and .2 including any material change in such particulars or estimate.

  .1 .3 The Contractor shall submit to the Architect the particulars and estimate referred to in clause 11.2.1.2.1 and .2 and the further notices referred to in clause 11.2.1.2.3 to the extent that such particulars and estimate have not been included in the notice given in accordance with clause 11.2.1.1 and shall, if so requested by the Sub-Contractor, join with the Sub-Contractor in requesting the consent of the Architect under clause 35.14 of the Main Contract Conditions.

The wording of the above sub-clauses is very similar to that contained in SF 80 when describing the action which the contractor should take. The contractor is obliged to submit to the architect the particulars and estimate referred to above and 'if so requested by the Sub-Contractor, join with the Sub-Contractor in

186

requesting the consent of the Architect under Clause 35.14 of the Main Contract Conditions'. This takes us back to SF 80 where we find:

35.14 .1 The Contractor shall not grant to any Nominated Sub-Contractor any extension of the period or periods within which the Sub-Contract Works (or where the Sub-Contract Works are to be completed in parts any part thereof) are to be completed except in accordance with the relevant provisions of Sub-Contract NSC/4 or NSC/4a as applicable which requires the written consent of the Architect to any such grant.

35.14 .2 The Architect shall operate the relevant provisions of Sub-Contract NSC/4 or NSC/4a as applicable upon receiving any notice particulars and estimate and a request from the Contractor and any Nominated Sub-Contractor for his written consent to an extension of the period or periods for the completion of the Sub-Contract Works or any part thereof as referred to in clause 11.2.2 of Sub-Contract NSC/4 or NSC/4a as applicable.

The reason for returning to the main contract conditions is that the sub-contract conditions cannot ignore the contract between employer and contractor which are set out in SF 80.

SF 80 requires the contractor *not* to grant any extension of time to the sub-contractor without the written consent of the architect (SF 80, cl. 35.14.1) but the architect is obliged to operate the relevant provision of NSC/4. This is a much closer linkage between the main and sub-contract conditions than existed between the 1963 edition of the Standard Form of Building Contract and the Green Form.

Our paper-chase now leads us back to NSC/4 to see the course of action which the architect must take.

We are, again, dependent upon the architect's 'opinion' and it is prudent to quote the sub-clauses verbatim:

11.2 .2 If on receipt of any notice, particulars and estimate under clause 11.2.1 and of a request by the Contractor and the Sub-Contractor for his consent under clause 35.14 of the Main Contract Conditions the Architect is of the opinion that:

.2 .1 any of the matters which are stated by the Sub-Contractor to be the cause of the delay is an act, omission or default of the Contractor, his servants or agents or his sub-contractors, their servants or agents (other than the Sub-Contractor, his servants or agents) or the occurrence of a Relevant Event; and

.2 .2 the completion of the Sub-Contract Works or any part thereof is likely to be delayed thereby beyond the period or periods stated in the Tender, Schedule 2, item 1C, or any such revised period or periods

then the Contractor shall, with the written consent of the Architect, give an extension of time by fixing such revised or further revised period or periods for the completion of the Sub-Contract Works or any part thereof as the Architect in his written consent then estimates to be fair and reasonable. The Contractor shall, in agreement with the Architect, when fixing such revised period or periods state:

187

It will be noted that in addition to the 'occurrence of a Relevant Event', clause 11.2.2.1 refers to 'an act, omission or default of the Contractor . . .' as a possible cause of delay.

In clause 11.2.2 we have the rather quaint, three-sided arrangement where it is the architect who considers the sub-contractor's notice, particulars and estimate which he, the architect, received via the contractor and which he (the architect again) must consent to (if he is of the opinion that he should do so) via the contractor. The principle has not changed very much since the 1963 edition but the ponderous procedure is more readily apparent in the new forms.

The remainder of the clause reads:

.2   .3   which of the matters including any of the Relevant Events, referred to in clause 11.2.2.1 they have taken into account; and

.2   .4   the extent, if any, to which the Architect, in giving his written consent, has had regard to any instruction requiring as a Variation the omission of any work issued under clause 13.2 of the Main Contract Conditions since the previous fixing of any such revised period or periods for the completion of the Sub-Contract Works or any part thereof,

and shall, if reasonably practicable having regard to the sufficiency of the aforesaid notice, particulars and estimate, fix such revised period or periods not later than 12 weeks from the receipt by the Contractor of the notice and of reasonably sufficient particulars and estimates, or, where the time between receipt thereof and the expiry of the period or periods for the completion of the Sub-Contract Works or any part thereof is less than 12 weeks, not later than the expiry of the aforesaid period or periods.

That completes the first exercise by the contractor of the duty which, as may be seen above, is similar to that performed by the architect in respect of the contractor's notices. We have now reached the point where the contractor, in agreement with the architect, has fixed revised periods and stated which matters have been taken into account and the extent to which the architect has had regard to any instructions he may have given requiring as a variation the omission of any work.

The contractor, it will be noted, is required to fix the revised period or periods not later than twelve weeks from his receipt of the notice, etc. from the sub-contractor or not later than the expiry of the period/s for the completion of the sub-contract works. This requirement may place the contractor in some difficulties because if experience from the past is to be used as a guide it is likely that both contractor and sub-contractor will be giving notices of delay at much the same times and probably in respect of a number of different relevant events. It may be difficult for the contractor to fix a revised period not later than twelve weeks from receipt of the sub-contractor's notice if that notice (together with particulars and estimate) has to be considered by the architect who, in turn, has twelve weeks to 'exercise his duty'.

Period for completion may be shortened

SF 80 appears to take into account the possibility of 'time' being important to the employer. There is, therefore, provision in clause 11.2.3 for the architect to issue an instruction which requires as a variation the *omission of work*. Presumably the thinking is that if there is less work to be done it will take less time and the period for completion of the sub-contract works may therefore be shortened. The arrangement in this respect is similar in both SF 80 and NSC/4. The complete text of clause 11.2.3 in NSC/4 is given below:

11.2 .3 After the first exercise by the Contractor of the duty under clause 11.2.2, the Contractor, with the written consent of the Architect, may fix a period or periods for completion of the Sub-Contract Works or any part thereof shorter than that previously fixed under clause 11.2.2 if, in the opinion of the Architect, the fixing of such shorter period or periods is fair and reasonable having regard to any instructions issued under clause 13.2 of the Main Contract Conditions requiring as a Variation the omission of any work where such issue is after the last occasion on which the Contractor with the consent of the Architect made a revision of the aforesaid period or periods.

There is an opportunity for 'second thoughts' on the part of the contractor and architect with regard to lengthening, shortening or confirming the period or periods for completion of the sub-contract works not later than the expiry of twelve weeks from the date of practical completion of the sub-contract works or from the date of practical completion of the main contract, whichever first occurs. Clauses 11.2.4.1–11.2.4.3 provide that:

11.2 .4 Not later than the expiry of 12 weeks from the date of practical completion of the Sub-Contract Works certified under clause 35.16 of the Main Contract Conditions or from the date of Practical Completion of the Main Contract, whichever first occurs, the Contractor with the written consent of the Architect shall either:

.1 fix such a period or periods for completion of the Sub-Contract Works or any part thereof longer than that previously fixed under clause 11.2 as the Architect in his written consent considers to be fair and reasonable having regard to any of the matters referred to in clause 11.2.2.1 whether upon reviewing a previous decision or otherwise and whether or not the matters referred to in clause 11.2.2.1 have been specifically notified by the Sub-Contractor under clause 11.2.1; or

.2 fix such a period or periods for completion of the Sub-Contract Works or any part thereof shorter than that previously fixed under clause 11.2 as the Architect in his written consent considers to be fair and reasonable having regard to any instruction issued under clause 13.2 of the Main Contract Conditions requiring as a Variation the omission of any work where such issue is after the last occasion on which the Contractor made a revision of the aforesaid period or periods; or

.3 confirm in writing to the Sub-Contractor the period or periods for the completion of the Sub-Contract Works previously fixed.

> Provided always the Sub-Contractor shall use constantly his best endeavours to prevent delay in the progress of the Sub-Contract Works, howsoever caused, and to prevent any such delay resulting in the completion of the Sub-Contract Works being delayed or further delayed beyond the period or periods for completion; and the Sub-Contractor shall do all that may reasonably be required to the satisfaction of the Architect and the Contractor to proceed with the Sub-Contract Works.

The provisos contained in the final paragraph, above, make demands of the sub-contractor similar to those which are made of the contractor in SF 80, clause 25.3.4.1.

Earlier in this chapter it was stated that the contractor and/or the sub-contractor require an extension of time because without it they may be liable to pay the employer liquidated damages. There is, however, a clearly defined list of events, named 'Relevant Events' for which an extension of time may be given. The relevant events are set out in clause 25. of SF 80 and in clause 11.25 of NSC/4. The lists are very similar and as we are concerned primarily with the sub-contractor's interests we will use the NSC/4 clause.

The clause begins:

11.2 .5 The following are the Relevant Events referred to in clause 11.2.2.1:

and the first relevant event is:

.5 .1 force majeure;

'Force majeure' is, literally, 'forces beyond one's control', and is a term used in legal documents to embrace a wide range of major events some of which are included as other relevant events in clause 11.2.5. Generally, however, acts of God, war, earthquakes and the like would come in the *force majeure* category.

The relevant event referred to in clause 11.2.5.2 reads:

.5 .2 exceptionally adverse weather conditions;

The key word is 'exceptionally'. Contractors and sub-contractors alike must accept the fact that adverse weather conditions are, in Britain, the rule rather than the exception and that to qualify as a relevant event the weather conditions must be *exceptionally* adverse. It is, in practice, difficult for a contractor or sub-contractor to establish when the adversity of the weather conditions becomes exceptional because so many factors are involved. It may be, for instance, that adverse weather conditions, however exceptionally adverse, would have little or no effect on a project which had advanced to a point where the building was weathertight. On the other hand weather conditions would not have to be even *exceptionally* adverse before a foundation engineering or piling sub-contractor working on a heavy clay soil found it impossible to carry out his work.

How then may one measure the point beyond which the weather conditions become exceptionally adverse?

One source of information on which measurements may be based is the Meteorological Office records. By comparison of records over a period of time it may be possible for the contractor or sub-contractor to determine exceptions from the norm which will satisfy the architect that the adverse weather conditions which have been experienced on the project under consideration have been exceptional and thus qualify as a relevant event. These records are an imprecise method of measurement because they do not normally take into account the time of the day or night when the conditions occurred nor do they take into account the nature of the site but they are, at least, an indication of condition and they may provide a starting point from which a more detailed 'case' may be prepared.

The relevant event referred to in clause 11.2.5.3 reads:

> .5 .3 loss or damage occasioned by any one or more of the 'Clause 22 Perils',

'Clause 22 Perils' are those against which the contractor must insure to comply with clause 22 of SF 80. They appear in Part 1 of SF 80 in the definitions and they make a formidible list which starts with fire, lighting, explosion, storm and includes overflowing water tanks, articles dropped from aircraft and civil commotion.

Loss or damage caused by ionizing radiations, contamination by radioactivity from any nuclear fuel, from nuclear waste, from pressure waves caused by super-sonic aircraft or other aerial devices and some other twentieth-century nastinesses are specifically excluded but these items may qualify for an extension of time as 'force majeure'. The majority of these perils would be unlikely to delay just the sub-contractor's work and in the unfortunate event of them occurring the appropriate action would almost certainly be taken by the contractor. Nevertheless, the sub-contractor should beware because one or two of the less catastropic items, such as overflowing water tanks, might delay sub-contract work without having an effect which the contractor regarded as sufficiently important for him to notify as a delay. An example of such a situation occurred in a building which was almost completed. Delicate sprays of water emerged from light-switches and the bowlshades below the ceiling more than half-filled with water. 'All we need is some gold-fish,' said the foreman electrician as he surveyed the ceiling fittings.

Water from an overflowing tank had percolated under the floor and gained access into some electrical conduits which it treated as purpose-made aquaducts. The conduit had captured almost all the water and adjoining works were virtually unaffected.

The relevant event referred to in clause 11.2.5.4 reads:

> .5 .4 civil commotion, local combination of workmen, strike or lock-out affecting any of the trades employed upon the Works or any of the trades engaged in the preparation, manufacture or transportation of any of the goods or materials required for the Works;

It is important for the sub-contractor to note that the clause refers not only to men employed upon the works but also to *trades engaged in the preparation, manufacture or transportation* of any of the goods or materials required for the works. It will almost certainly be necessary for the sub-contractor to obtain appropriate information relating to off-site strikes in order to demonstrate to the architect and contractor the effect of such strikes on the project as a whole.

The relevant event referred to in clause 11.2.5.5 reads:

.5 .5 compliance by the Contractor and/or Sub-Contractor with the Architect's instructions:

.5 .5 .1 under clauses 2.3, 13.2, 13.3, 23.2, 34, 35 or 36 of the Main Contract Conditions, or

.5 .5 .2 in regard to the opening up for inspection of any work covered up or the testing of any of the work, materials or goods in accordance with clause 8.3 of the Main Contract Conditions (including making good in consequence of such opening up or testing) unless the inspection or test showed that the work, materials or goods were not in accordance with the Main Contract or the Sub-Contract as the case may be;

The clauses referred to in clause 11.2.5.5.1 are concerned with:

(a) clause 2.3, restrictions in contracts of sale, etc. – limitations of liability of sub-contractor in the event of him being required to enter in a sub-sub-contract;

(b) clauses 13.2 and 13.3 – disturbance of regular progress of sub-contract works by an act, omission or default of the contractor, another sub-contractor or any sub-sub-contractor;

(c) clause 23.2 – contractor's right to set off against the sub-contractor;

(d) clauses 34, 35 and 36, choice of fluctuation provisions.

These relevant events should not be overlooked even though they appear only as clause numbers.

It is apparent from clause 11.2.5.5.2 that the contractor may *not* regard 'the opening up for inspection of any work' as a relevant event if the inspection or test which the architect instructed should be made showed that the work, etc. was not in accordance with the main contract or the sub-contract as the case may be. If, however, a sub-contractor has been delayed as a result of the contractor's failure in this respect he, the sub-contractor, should look to clause 11.2.2.1 which requires the contractor (no doubt unenthusiastically) with the written consent of the architect to give the sub-contractor an extension of time.

In this instance the cause of the delay is not a relevant event but 'an act, omission or default of the Contractor'. It is suggested that the contractor may lack enthusiasm because the sub-contractor may have grounds for a claim against the contractor in addition to obtaining an extension of time as a result of the contractor's act, omission or default.

The next relevant event which is set out in clause 11.2.5.6 and which reads:

.5 .6 the Contractor, or the Sub-Contractor through the Contractor, not having received in due time necessary instructions, drawings, details or levels from the Architect for which the Contractor or the Sub-Contractor, through the Contractor, specifically applied in writing provided that such application was made on a date which having regard to the Completion Date or the period or periods for the completion of the Sub-Contract Works was neither unreasonably distant from nor unreasonably close to the date on which it was necessary for the Contractor or the Sub-Contractor to receive the same;

A most important aspect of this relevant event, frequently overlooked by contractors and sub-contractors, is that they cannot just sit back and blame the architect and expect him to give them extensions of time because they have not received in due time necessary instructions, etc. *unless* (and it is an important 'unless') they have specifically applied in writing for the instructions, etc. on a date which having regard to the completion date, etc. was neither unreasonably distant from nor unreasonably close to the date on which it was necessary for them, the contractors and sub-contractors, to receive them. At what point is a date unreasonably distant from or close to?

Provided the sub-contractor is able to explain, if asked by the contractor or architect, why he needs the instructions, etc. by a certain date, it is most unlikely that his application would be regarded as unreasonable. It is the sub-contractor who has to carry out and complete the sub-contract works and only he is able to say when it will be necessary for him to place orders for this or that piece of equipment or commence off-site manufacture of part of the sub-contract work. The intention of restricting the timing of the applications for instructions, etc. appears to be to assist the architect or design team to plan their design work and order their priorities.

There has been considerable discussion about whether a contractor or sub-contractor would be considered to meet the timing requirement if, when he entered into contract, he were immediately to provide a schedule of dates, covering a period some months ahead showing when he required instructions, etc. for various aspects of his work. For practical purposes the instruction dates may be more conveniently presented in the form of a bar-chart which relates the dates by which instructions should be given to the dates when the work is to be carried out (see Fig. 13.1.). A photocopy of the relevant part of the master plan may be used for this purpose. Is his application, which is being made far in advance of the time when he actually requires the drawings, etc., 'unreasonably distant from'?

It will be noted from the wording of the clause that it is the date of the *application* which has the time restriction.

When considering points of contract and procedures there is something to be said for putting oneself in the position of an arbitrator who has been called upon to make an award. Would he have to take into account the contractual as well as the practical aspects? The answer is, of course, that he most certainly would. The sub-contractor would, then, be well advised to be on the safe side – in contract matters wearing belt and braces even with self-supporting trousers is a sensible precaution. He will make his own plans when he enters his contract so he should acquaint the contractor with his requirements at that time in the form of an application. The sub-contractor's schedule may then provide the basis for subsequent 'applications' as the project progresses. If the works are delayed, the

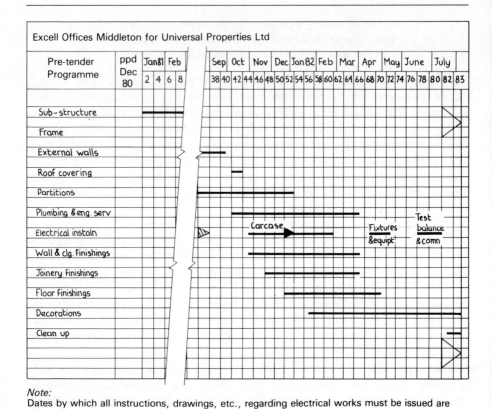

Fig. 13.1 Application for instructions, drawings, etc.

sub-contractor should advise the contractor that the actual dates by which instructions should be given may be varied from those shown on the schedule, provided that the times between instructions and commencements on site are maintained as shown on the schedule. Alternatively, the sub-contractor may have committed a sub-sub-contractor to keep a certain period of time free in his workshop, in which event it would be reasonable for the sub-contractor to insist that the architect adheres to the *date* shown on the schedule.

Clause 11.2.5.6 is important to the sub-contractor for two reasons. The first is that the relevant event to which it refers may provide him with an extension of time, and the second that such an extension of time may lead to reimbursement by reason of the regular progress of the sub-contract works being materially affected.

We will discuss the second matter later. Relevant event 11.2.5.7 reads:

> .5 .7 delay on the part of Nominated Sub-Contractors (other than the Sub-Contractor) or of Nominated Suppliers in respect of the Works which the Contractor has taken all practicable steps to avoid or reduce;

A sub-contractor is very vulnerable. It is true that a contractor has only limited powers to control the progress of sub-contractor's sub-contract works but a sub-contractor cannot in any way control the progress of either the contractor's or other sub-contractor's works. Furthermore, a sub-contractor is, as may be seen from the wording of clause 11.2.5.7, dependent upon the contractor having 'taken all practicable steps to avoid or reduce' delay on the part of all nominated sub-contractors if he, our sub-contractor, is to be eligible for an extension of time.

There is really little that the sub-contractor can do in this respect, except ensure that *he* does not cause delay and give his written notice 'forthwith' (as required by cl. 11.2.1.1) in the event of him being delayed.

Relevant event 11.2.5.8 reads:

.5 .8 .1 the execution of work not forming part of the Main Contract by the Employer himself or by persons employed or otherwise engaged by the Employer as referred to in clause 29 of the Main Contract Conditions or the failure to execute such work;

.5 .8 .2 the supply by the Employer of materials and goods which the Employer has agreed to provide for the Works or the failure so to supply;

From the sub-contractor's standpoint this clause has a similar effect to the previous clause. The work '*not forming part of the Main Contract*' which is carried out 'by the Employer himself or by persons employed or otherwise engaged by the employer', referred to above (my italics), replaces the 'artists and tradesmen' for whom the employer is responsible (but who are *not* nominated sub-contractors) who are referred to in clause 29 of the 1963 edition of the Standard Form of Building Contract. There is no equivalent in the 1963 edition to the 'supply by the Employer of materials and goods which the Employer has agreed to provide for the works or the failure so to supply' which is referred to in clause 11.5.8.2 above.

Relevant event 11.2.5.9 reads:

.5 .9 the exercise after the Date of Tender by the United Kingdom Government of any statutory power which directly affects the execution of the Works by restricting the availability or use of labour which is essential to the proper carrying out of the Works, or preventing the Contractor or Sub-Contractor from, or delaying the Contractor or Sub-Contractor, in, securing such goods or materials or such fuel or energy as are essential to the proper carrying out of the Works;

There is no direct equivalent of this clause in earlier editions of the Standard Form of Building Contract. It would appear that there is usually no action which the sub-contractor need take in this event because the contractor would, generally, be affected by the government's action and should take appropriate action as required by the conditions of contract. Nevertheless, it is conceivable that the government's action *may* affect only the work of a specialist sub-contractor and for this reason the sub-contractor should give his written notice. But even if the delay which the government caused were of an apparently general nature, the sub-contractor should remember that an extension of time arises from a 'written notice' and act accordingly.

Relevant event 11.2.5.10 is concerned with the contractor's or sub-contractor's *inability for reasons beyond his control* to secure labour and/or materials which are essential to the proper carrying out of the works:

> .5 .10 .1 the Contractor's or Sub-Contractor's inability for reasons beyond his control and which he could not reasonably have foreseen at the Date of Tender for the purposes of the Main Contract or the Sub-Contract as the case may be to secure such labour as is essential to the proper carrying out of the Works:

> .5 .10 .2 the Contractor's or Sub-Contractor's inability for reasons beyond his control and which he could not reasonably have foreseen at the Date of Tender for the purposes of the Main Contract or the Sub-Contract as the case may be to secure such goods or materials as are essential to the proper carrying out of the Works;

The sub-contractor should take note of the wording in italics, above. This is another instance where the sub-contractor's work may be delayed by the inability of another party (the contractor in this instance) properly to carry out the works.

Local authorities occasionally seek to delete this clause (or rather the equivalent clause in the main contract) from their contract with the contractor. The sub-contractor would be wise to ensure that he does not agree to the deletion of clause 11.2.5.10. He should, also, satisfy himself of the contractor's undertaking with the employer by referring to the main contract conditions.
Relevant event 11.2.5.11 reads:

> .5 .11 the carrying out by a local authority or statutory undertaker of work in pursuance of its statutory obligations in relation to the Works, or the failure to carry out such work;

Wording very similar to that given above occurs in clause 23(1) of the 1963 edition as revised in 1975. By 1975 the performance of local authorities and statutory undertakers had delayed so many contracts that the JCT considered it necessary to incorporate this clause. Statutory undertakers are the various 'boards' which have been formed to provide electricity, gas, water, etc. The contractor and sub-contractor are entitled to an extension of time in respect of clause 11.2.5.11 *only* if the delay is caused by the local authority or statutory undertaker when executing its statutory obligations. If they are acting as nominated sub-contractors, as would be electricity boards which undertake electrical installation works, an application for an extension of time by a contractor or sub-contractor should be under the clause 11.2.5.7 – the relevant event which refers to nominated sub-contractors. Here, again, there is little that our sub-contractor can do except ensure that he gives his notice of delay promptly.

The relevant event referred to in clause 11.2.5.12 is an innovation introduced in the 1980 edition of the Standard Form of Building Contract. The clause reads:

11.2 .5 .12   failure of the Employer to give in due time ingress to or egress from the site of the Works or any part thereof through or over any land, buildings, way or passage adjoining or connected with the site and in the possession and control of the Employer, in accordance with the Contract Bills and/or the Contract Drawings, after receipt by the Architect of such notice, if any, as the Contractor is required to give, or failure of the Employer to give such ingress or egress as otherwise agreed between the Architect and the Contractor;

This is another relevant event which would in all probability concern the contract or more than the nominated sub-contractor and for this reason the sub-contractor would not expect to take action in this matter. It is, however, possible that a sub-contractor may be denied access to, say, part of an existing building which was being renovated in stages, etc. and to which it was important that he had access in order to work. In this event the contractor may not be so directly concerned and the sub-contractor would have to take the initiative to give his written notice to the contractor.

The final relevant event in clause 11 – clause 11.2.5.13 – concerns the sub-contractor's right to suspend the sub-contract works as a result of the contractor failing 'to discharge his obligation to make any payment to the sub-contractor' in accordance with clause 21.8. The wording of the relevant event is:

.5 .13   the valid exercise by the Sub-Contractor of the right in clause 21.8 to suspend the further execution of the Sub-Contract Works.

from which it is apparent that if the sub-contractor is to be eligible for an extension of time the exercise of his right in accordance with clause 21.8 must be 'valid'.

Having examined clause 11 of NSC/4 in some detail we may consider the action which the sub-contractor should take to comply with that clause whenever it becomes reasonably apparent that the 'commencement, progress or completion of the sub-contract works or any part thereof is being or is likely to be delayed' (NSC/4, cl. 11.2.1.1).

A case study

To illustrate forms of words which may be used in a letter giving notice let us assume that the Excell Offices project has reached the end of January 1982 (see Fig. 13.1) and several events have occurred.

There is, of course, an anomaly in that it is not January 1982 but this need not bother us. The lapse of time always presents problems on case studies. The events which have occurred are:

(a)   information regarding some fixtures and equipment which, as may be seen in Fig. 13.1, was stated must be issued by the end of week 52 (four weeks ago) has not been issued;

(b)   the progress of the main contractor's work is behind programme on the third and fourth floors so that the sub-contract works cannot proceed. This failure became apparent some three months ago and the contractor has already been notified;

(c) ceiling fittings for the offices which are items which were specified at time of tender have not been delivered by the sub-contractor's supplier and there are strong rumours that he is in financial difficulties and likely to cease trading. The sub-contractor has only recently heard these rumours;

(d) there was a fire in the basement three days ago caused by the contractor's plumber which has seriously damaged the equipment in the switchroom which was almost ready for commissioning.

These events are by no means an exaggeration of real life. In practice there is seldom just one event which causes delay.

The sub-contractor must satisfy himself that the events are 'relevant', as listed in clause 11.2.5, or that the contractor is at fault (see cl. 11.2.2.1).

Using semi-standard letters

A specialist contractor who regularly undertakes work as a nominated sub-contractor on building projects should consider adopting a form of words which he may use on all projects when a delay occurs. Only the actual events change so the use of a semi-standard format saves the sub-contractor from having to 'watch his words' every time he writes a letter.

Table 13.1 demonstrates a possible format for a letter, the 'standard' wording is in italics and the 'unique' wording, applicable to events (a) to (d), is in typeface in the following example. The words in brackets marked with an asterisk are alternatives and only the applicable alternative would be included in an actual letter. The clause numbers are intended to indicate the applicable clause in NSC/4 and need not be included in the letter.

It will be noted that the letter is concerned with both the initial written notice of a delay and the 'further' notice. It is uncertain if the final paragraph is of any great significance but it is a statement of the sub-contractor's intention which appears to be supported by his deeds.

The wording in (a) to (c), in the letter, follows as closely as possible that used in the NSC/4 clauses.

It will also be noted that *the sub-contractor merely notifies delay* – it is the contractor's responsibility, in agreement with the architect, to give an extension of time. A word processor may be useful for such semi-standard letters.

Arbitration

This, then, should be all that the sub-contractor has to do to comply with NSC/4 clause 11 but if he feels aggrieved by the architect's decision or by his failure to give a decision the sub-contractor may, with the agreement of the contractor, use the contractor's name and if necessary join with him in arbitration proceedings (cl. 11.3).

Table 13.1 Form of words for 'Notice of Delay'

The Contractor
re: Excell Offices, Middleton

We give notice that it has become apparent that the (commencement/progress/completion*) of (part of*) the sub-contract works is (being/likely to be*) delayed by:
(a) us not having received in due time necessary instructions from the architect regarding fixtures and equipment for which, through you, we specifically applied in writing on 19 June 1981 (11.2.5.6);
(b) our inability, for reasons beyond our control and which we could not reasonably have foreseen at the date of tender, to secure ceiling fittings for the offices (11.2.5.10.2);
(c) damage occasioned by fire in the basement during the night of 25–26 January (11.2.5.3).

The expected effect of and the estimated extent of the expected delay in the completion of the sub-contract works in respect of the relevant events is as follows:
  (i) the delivery period for the fixtures and equipment is ten weeks from the date of our order and we must be in receipt of instructions by 31 January if no delay is to be caused. We cannot estimate the extent of the expected delay until we receive the instructions;
 (ii) it has become necessary for us to place an order for the ceiling fittings to be fixed in the offices with another supplier who cannot deliver the fittings for twelve weeks. The absence of the fittings will not prevent all adjacent work being carried out but we estimate a delay of five working days;
(iii) we understand from your letter of yesterday's date that we must take no action to re-instate the damage caused by the fire until we receive your instructions. Certain replacement equipment for which there is a fourteen weeks delivery period will have to be ordered and we have fifteen working days work in the switchroom. We estimate the extent of the delay in the completion of our works in the basement as seventeen weeks from the time when we receive your instructions to proceed.

We also give you notice, further to our letter dated 25 October 1981 that the progress of your work in the third and fourth floor office areas delayed the completion of our work in these areas by five working days, not ten working days as we estimated when we wrote to you in October (11.2.1.2.3).

We will use constantly our best endeavours to prevent delay in the progress of the sub-contract works.
Signed ....

It is important for the sub-contractor to remember that the above letter should lead to an extension of the sub-contract time but *it does not lead to him obtaining reimbursement for any direct loss and/or expense* caused to him. This is a matter dealt with in clause 13, discussed below.

Failure to complete on time

We have not considered the effects of the failure of the sub-contractor to complete on time. This eventuality is the subject of NSC/4 clause 12 which reads:

12.1 If the Sub-Contractor fails to complete the Sub-Contract Works (or any part thereof) within the period or periods for completion or any revised period or periods as provided in clause 11.2.2, the Contractor shall so notify the Architect and give to the Sub-Contractor a copy of such notification.

12.2 The Sub-Contractor shall pay or allow to the Contractor a sum equivalent to any loss or damage suffered or incurred by the Contractor and caused by the failure of the Sub-Contractor as aforesaid. Provided that the Contractor shall not be entitled so to claim unless the Architect in accordance with clause 35.15 of the Main Contract Conditions shall have issued to the Contractor (with a copy to the Sub-Contractor) a certificate in writing certifying any failure notified under clause 12.1.

The second paragraph, clause 12.2, leads to clause 35.15 in SF 80. We established, above, that *'the Contractor shall not grant to any Nominated Sub-Contractor any extension of the period or periods within which the Sub-Contract Works are to be completed . . .'* (SF 80 cl. 35.14.1) and the architect must be satisfied that this clause has been properly applied before SF 80 cl. 15, referred to in NSC/4 cl.12 above, comes into effect.

Certificate of failure

SF 80 cl. 35.15 is concerned with failure of the nominated sub-contractor to complete his works. It reads:

35.15 .1 If any Nominated Sub-Contractor fails to complete the Sub-Contract Works (or where the Sub-Contract Works are to be completed in parts any part thereof) within the period specified in the Sub-Contract or within any extended time granted by the Contractor with the written consent of the Architect and the Contractor so notifies the Architect with a copy to the Nominated Sub-Contractor, then, provided that the Architect is satisfied that clause 35.14 has been properly applied, the Architect shall so certify in writing to the Contractor. Immediately upon the issue of such a certificate the Architect shall send a duplicate thereof to the Nominated Sub-Contractor.

35.15 .2 The certificate of the Architect under clause 35.15.1 shall be issued not later than 2 months from the date of notification to the Architect that the Nominated Sub-Contractor has failed to complete the Sub-Contract Works or any part thereof.

It will be noted that the architect is required to *issue a certificate* that the nominated sub-contractor has failed to complete the sub-contract works, if he has so failed.

This clause is the equivalent of the arrangement contained in clause 27(d)(ii) of the 1963 edition of the Standard Form of Building Contract. Its importance to both contractor and sub-contractor is that it enables the contractor to claim against the sub-contractor for direct loss and/or expense which he may have suffered as a result of the sub-contractor's failure.

Experience with the 1963 edition has shown that architects are generally reluctant to issue a certificate that the nominated sub-contractor has failed to complete

200

the sub-contractor works. Sub-contractors should hope that this reluctance will continue because once a contractor is armed with a certificate he is in quite a strong position *vis-à-vis* the sub-contractor.

Having considered the effect of the delays and extensions of time which all too frequently occur on building projects, it is possible to consider 'matters affecting regular progress – direct loss and/or expense – Contractor's and Sub-Contractor's rights'.

The primary concern is with the sub-contractor's rights and obligations but it is necessary first to consider the relevant clause in SF 80 (as was the case when considering delays and extensions of time) because of the interrelationship of SF 80 and NSC/4.

Contractor makes application; architect ascertains loss

Clause 26 is the relevant clause in SF 80, the introductory paragraph of which reads:

26   Loss and expense caused by matters materially affecting regular progress of the Works

26.1   If the Contractor makes written application to the Architect stating that he has incurred or is likely to incur direct loss and/or expense in the execution of this Contract for which he would not be reimbursed by a payment under any other provision in this Contract because the regular progress of the Works or of any part thereof has been or is likely to be materially affected by any one or more of the matters referred to in clause 26.2; and if and as soon as the Architect is of the opinion that the regular progress of the Works or of any part thereof has been or is likely to be so materially affected as set out in the application of the Contractor then the Architect from time to time thereafter shall ascertain, or shall instruct the Quantity Surveyor to ascertain, the amount of such loss and/or expense which has been or is being incurred by the Contractor; provided always that: . . .

Ignoring for a moment the three concluding words, what is the effect of the above paragraph?

The gist of clause 26.1 first appeared in the 1963 edition of the Standard Form of Building Contract (cl. 24) and SF 80 does not change the 1963 edition wording very greatly.

The 'written application' is what is frequently described as a 'claim' but that word is not used in SF 80, clause 26 nor in NSC/4, clause 13, the sub-contract counterpart, in this context.

The word 'claim' was used in the old green form even though it did not appear in the 1963 edition of the main contract form. The avoidance of the word in SF 80 and NSC/4 is another example of the 1980 'coming together' of main and sub-contract forms. Sub-contractors familiar with the Green Form will, then, have to change their thinking in some respect. As far as SF 80, clause 26.1 is concerned the position is simply that the contractor makes a written application

201

stating that 'he has incurred or is likely to incur direct loss and/or expense . . . (etc.)' and provided the architect 'is of the opinion that the regular progress of the works . . . is likely to be so materially affected as set out in the application' he shall ascertain the amount of such loss and/or expense and (and here we move forward to cl. 26.5):

26.5　Any amount from time to time ascertained under clause 26 shall be added to the Contract Sum.

which is self explanatory.

The provisions referred to in the last three words of clause 26.1 are that the contractor's application shall be timely, that he shall submit any information and details of such loss and/or expense as the architect or quantity surveyor requests. These provisions are essentially the same as those which apply to the sub-contractor so they will be discussed in more detail when considering NSC/4, clause 13. For the same reason the 'matters referred to in clause 26.2' (see cl. 26.1 above) will be left for the time being.

It is interesting that although the architect is responsible for ascertaining the loss and/or expense he may ask the contractor to provide him, or the quantity surveyor, with the details he needs to do so.

Relevance of extensions of completion date

A clause in SF 80 which is not echoed in NSC/4 in clause 26.3 which reads:

26.3　If and to the extent that it is necessary for ascertainment under clause 26.1 of loss and/or expense the Architect shall state in writing to the Contractor what extension of time, if any, has been made under clause 25 in respect of the Relevant Event or Events referred to in clause 25.4.5.1 (so far as that clause refers to clauses 2.3, 13.2, 13.3 and 23.2) and in clauses 25.4.5.2, 25.4.6, 25.4.8 and 25.4.12.

The clause is self-explanatory and the relevant events and events referred to in clause 25.4.5.1 are of little significance to the sub-contractor.

Contractor shall pass on sub-contractor's application

Of considerable importance to the sub-contractor is clause 26.4.1:

26.4　.1　The Contractor upon receipt of a written application properly made by a Nominated Sub-Contractor under clause 13.1 of Sub-Contract NSC/4 or NSC/4a as applicable shall pass to the Architect a copy of that written application. If and as soon as the Architect is of the opinion that the regular progress of the Sub-Contract Works or of any part thereof has been or is likely to be materially affected as referred to in clause 13.1 of Sub-Contract NSC/4 or NSC/4a and as set out in the application of the Nominated Sub-Contractor then the Architect shall himself ascertain, or shall instruct the

Quantity Surveyor to ascertain, the amount of such loss and/or expense.

This clause obliges the contractor to pass on to the architect any written applications from the sub-contractor and obliges the architect, provided he is of the right opinion, etc. to ascertain the amount of such loss and/or expense which, as referred to in clause 26.5 above, shall be added to the contract sum.

Finally, as far as SF 80, clause 26 is concerned, clause 26.4.2 obliges the architect to inform the contractor and sub-contractor of the length of the revision of the period or periods for completion of sub-contract works:

26.4 .2 If and to the extent that it is necessary for the ascertainment of such loss and/or expense the Architect shall state in writing to the Contractor with a copy to the Nominated Sub-Contractor concerned what was the length of the revision of the period or periods for completion of the Sub-Contract Works or of any part thereof to which he gave consent in respect of the Relevant Event or Events set out in clause 11.2.3.5.1 (so far as that clause refers to clauses 2.3, 13.2, 13.3 and 23.2 of the Main Contract Conditions), 11.2.3.5.2, 11.2.3.6, 11.2.3.8 and 11.2.3.12 of Sub-Contract NSC/4 or NSC/4a as applicable.

Sub-contractor makes application to contractor and architect ascertains loss. The sub-contractor's position is set out in NSC/4, clauses 13.1.1 and 13.1.3 which read:

13 Matters affecting regular progress – direct loss and/or expense – Contractor's and Sub-Contractor's rights

13.1 .1 If the Sub-Contractor makes written application to the Contractor stating that he has incurred or is likely to incur direct loss and/or expense in the execution of the Sub-Contract for which he would not be reimbursed by a payment under any other provision in the Sub-Contract by reason of the regular progress of the Sub-Contract Works or of any part thereof having been or being likely to be materially affected by any one or more of the matters referred to in clause 13.1.2 the Contractor shall require the Architect to operate clause 26.4 of the Main Contract Conditions so that the amount of that direct loss and/or expense, if any, may be ascertained. (Provided always that: . . .)

.1 .3 The Sub-Contractor shall submit to the Contractor such details of such loss and/or expense as the Contractor is requested by the Architect or the Quantity Surveyor to obtain from the Sub-Contractor in order reasonably to enable the ascertainment of that loss and/or expense under clause 26.4 of the Main Contract Conditions.

The above clauses are similar in many respects to the introductory paragraph of SF 80, clause 26.

The provisions referred to in the last three words of clause 13.1.1 are:

13.1 .1 .1 The Sub-Contractor's application shall be made as soon as it has become, or should reasonably have become, apparent to him that the regular progress of the Sub-Contract Works or of any part thereof has been or was likely to be affected as aforesaid; and

.1 .2 The Sub-Contractor shall submit to the Contractor such information in support of his application as the Contractor is requested by the Architect to obtain from the Sub-Contractor in order reasonably to enable the Architect to operate clause 26.4 of the Main Contract Conditions; and . . .

None of these provisions should present the sub-contractor with any difficulty. Clause 13.1.1 above, refers to matters for which the sub-contractor 'would not be reimbursed by a payment under any other provision of the sub-contract' which suggests that there are matters for which he *would* be reimbursed. These are variations, which are dealt with in clauses 16 and 17 and which are discussed in Chapter 12.

A list of matters

The matters affecting the regular progress of the sub-contract works for which the sub-contractor would be reimbursed are listed in clause 13.1.2. The list reads:

13.1 .2 The following are the matters referred to in clause 13.1.1:

.2.1 the Contractor, or the Sub-Contractor through the Contractor, not having received in due time necessary instructions, drawings, details or levels from the Architect for which the Contractor, or the Sub-Contractor through the Contractor, specifically applied in writing provided that such application was made on a date which having regard to the Completion Date or the period or periods for completion of the Sub-Contract Works was neither unreasonably distant from nor unreasonably close to the date on which it was necessary for the Contractor or the Sub-Contractor to receive the same; or

.2.2 the opening up for inspection of any work covered up or the testing of any of the work, materials or goods in accordance with clause 8.3 of the Main Contract Conditions (including making good in consequence of such opening up or testing), unless the inspection or test showed that the work, materials or goods were not in accordance with the Main Contract or the Sub-Contract as the case may be; or

.2.3 any discrepancy in or divergence between the Contract Drawings and/or the Contract Bills; or

.2.4 the execution of work not forming part of the Main Contract by the Employer himself or by persons employed or otherwise engaged by the Employer as referred to in clause 29 of the Main Contract Conditions; or the failure to execute such work or the supply by the Employer of materials and goods which the Employer has agreed to provide for the Works or the failure so to supply; or

.2.5 Architect's instructions issued in regard to the postponement of any work to be executed under the provisions of the Main Contract or the Sub-Contract; or

.2.6 failure of the Employer to give in due time ingress to or egress from the site of the Works, or any part thereof through or over any land, buildings, way or passage adjoining or connected with the site and in the possession and control of the Employer, in accordance with the Contract Bills and/or the Contract Drawings, after receipt by the Architect of such notice, if any, as the Contractor is required to give or failure of the Employer to give such ingress or egress as otherwise agreed between the Architect and the Contractor; or

.2.7 Architect's instructions issued under clause 13.2 of the Main Contract Conditions requiring a Variation or under clause 13.3 of the Main Contract Conditions in regard to the expenditure of provisional sums (other than work to which clause 13.4.2 of the Main Conditions refers).

Not all 'Relevant Events' are 'matters' leading to an entitlement to recover loss and/or expense.

These 'matters' are very similar to some of the 'relevant events' contained in clause 11.2.5. The significance of the two words in inverted commas, above, is that by no means all of the relevant events qualify for reimbursement.

When delays occur and the regular progress of the works has been materially affected there are frequently several concurrent causes. It is important for a sub-contractor to obtain an extension of time but whilst this may relieve him of the possibility of a claim against him by the contractor it will not reimburse him for loss and/or expense. If, then, the sub-contractor is faced with giving written notice to the contractor of more than one relevant event which is causing delay the sub-contractor may consider it prudent to concentrate on the relevant event/s which is included in the list of matters in clause 13.1.2 rather than the event/s which are not. There is, however, a risk which the sub-contractor should have in mind and this is that if he is not given an extension of time in respect of the item/s of which he has given notice he may find it difficult to introduce a new item at a later date.

Sub-contractor cooperates with contractor

It is difficult to contemplate reasons why the sub-contractor would be reluctant to comply with clause 13.1.4:

13.1.4 The Sub-Contractor shall comply with all directions of the Contractor which are reasonably necessary to enable the ascertainment which results from the operation of clause 13.1.1 to be carried out.

So far, we have been concerned with matters which lead to an application for reimbursement being made, albeit indirectly, to the employer. It is by no means unusual for the contractor to cause delays leading to disturbance of the regular progress of the sub-contractor's work.

Sub-contractor delayed by contractor

Provision for this eventuality is included in clause 13.2:

13.2 If the regular progress of the Sub-Contract Works (including any part thereof

205

which is sub-sub-contracted) is materially affected by any act, omission or default of the Contractor, his servants or agents, or any sub-contractor, his servants or agents or sub-sub-contractor (other than the Sub-Contractor, his servants or agents or sub-sub-contractors) employed by the Contractor on the Works, the Sub-Contractor shall within a reasonable time of such material effect becoming apparent give written notice thereof to the Contractor and the agreed amount of any direct loss and/or expense thereby caused to the Sub-Contractor shall be recoverable from the Contractor as a debt.

A sub-contractor will not usually find the contractor as cooperative in regard to a notice under clause 13.2 as under clause 13.1.2. The reasons are obvious. The sub-contractor and contractor will usually have common ground under clause 13.1.2 and be united in their application to the employer. Under clause 13.2 it will be the contractor who must pay if the sub-contractor's application is successful.

Contractor delayed by sub-contractor

The other side of the coin is disturbance of the regular progress of the works by the sub-contractor or by his sub-sub-contractor. Clause 13.3 refers to this:

13.3   If the regular progress of the works (including any part thereof which is sub-contracted) is materially affected by any act, omission or default of the Sub-Contractor, his servants or agents, or any sub-sub-contractor employed by the Sub-Contractor on the Sub-Contract Works, the Contractor shall within a reasonable time of such material effect becoming apparent give written notice thereof to the Sub-Contractor and the agreed amount of any direct loss and/or expense caused to the Contractor (whether suffered or incurred by the Contractor or by sub-contractors employed by the Contractor on the Main Contract Works from whom claims under similar provisions in the relevant sub-contracts have been agreed by the Contractor, sub-contractor and the Sub-Contractor) may be deducted from any monies due or to become due to the Sub-Contractor or may be recoverable from the Sub-Contractor as a debt.

This sub-clause is clear in its meaning and it is interesting that the word 'claim' is used.

Set-off safeguards

The sub-contractor is protected, to a large extent, from unreasonable deductions by the contractor from 'monies due or to become due to the sub-contractor' by clause 24 which describes the action which the sub-contractor may take if he disagrees with any set-off the contractor may be inclined to make. This clause is discussed in Chapter 14.

The final sub-clause in clause 13, clause 13.4, reserves for both the contractor and the sub-contractor 'any other rights or remedies which [they] may possess'.

It is important for the sub-contractor to remember that he must make 'written application to the Contractor'. This action is quite distinct from any 'written notice' of delay required by clause 11. This is not to say that the two tasks may not be combined in a single letter but it must be made clear that both 'notice' and 'application' are involved.

Table 13.1 shows a form of words for a semi-standard letter which may be used to give 'notice of delay'. A similar letter should be drafted for an 'application for re-imbursement'.

Chapter 14

Disputes, contractor's right to set-off, adjudication and arbitration

Whilst on many building projects there are differences of opinion and disagreements between the parties to the contracts, the majority are resolved more or less amicably and without the need to refer the matter to a third party for judicial settlement.

There are, however, occasions when the differences of opinion and disagreements pass beyond the point where the parties to a contract feel that they are able to reach a mutually acceptable settlement and decide that reference must be made to an independent third party.

The courses open to the parties to the dispute are litigation and arbitration.

Litigation

Litigation is 'the action or process of carrying on a suit in law or equity; *In litigation*: in process of investigation before a court of law' (OED). Litigation is the normal course of action for disputants but the JCT standard forms of building contracts provide for another means; arbitration.

Arbitration under NSC/4

To arbitrate is 'to examine, give judgement' (OED) and NSC/4, article 3 states the procedure to be adopted by the contractor and sub-contractor in the event of dispute:

3.1   In the event of any dispute or difference between the Contractor and Sub-Contractor, whether arising during the execution or after the completion or abandonment of the Sub-Contract Works or after the determination of the employment of Sub-Contractor under Sub-Contract NSC/4 (whether by breach or in any other manner), in regard to any matter or thing of whatsoever nature arising out of the Sub-Contractor in connection therewith, then either party shall give to the other notice in writing of such dispute or difference and such dispute or difference shall be and is hereby referred to the arbitration and final decision of a person to be agreed between the parties, or, failing such agreement within 14 days after either party has given to the other a written request to concur in the appointment of an Arbitrator, a person to be appointed on the request of either party by the President or a Vice-President for the time being of the Royal Institution of Chartered Surveyors or, alternatively, if the party requiring the appointment so decides, by the President or a Vice-President for the time being of the Royal Institute of British Architects.

This article binds the contractor and sub-contractor to refer disputes to arbitration. It excludes either party taking legal action against the other in respect of the sub-contract.

The procedure is that when a dispute is apparent, one party may give notice in writing to the other that the matter in dispute should be referred to arbitration.

Such a 'notice' may be in the form of a letter. The parties may choose their own arbitrator or ask for one to be appointed.

One of the major advantages claimed for arbitration is that the arbitrator should have technical skill or professional experience which makes him a suitable person to *examine* and *give judgement* on the matter in dispute so the disputants should make certain that the arbitrator meets their special needs.

Article 3.1, above, refers to the contractor and sub-contractor, the parties to the sub-contract, being involved in a dispute but there may well be other parties involved. Article 3.2 makes provision for this eventuality but there is also provision in clause 3.2.2 for the contractor or sub-contractor to require a different arbitrator to be appointed if they consider the arbitrator appointed is not appropriately qualified for their requirements:

3.2 .1 Provided that if the dispute or difference to be referred to arbitration under this Sub-Contract raises issues which are substantially the same as or connected with issues raised in a related dispute between
the Contractor and the Employer under the Main Contract
or
the Sub-Contractor and the Employer under Agreement NSC/2 or NSC/2a as applicable
or
the Contractor and any other nominated sub-contractor under Sub-Contract NSC/4 or NSC/4a as applicable
or
the Contractor and any Nominated Supplier whose contract of sale with the Contractor provides for the matters referred to in clause 36.4.8 of the Main Contract Conditions
and if the related dispute has already been referred for determination to an Arbitrator, the Contractor and the Sub-Contractor hereby agree that the dispute or difference under this Sub-Contract shall be referred to the Arbitrator appointed to determine the related dispute; and such Arbitrator shall have power to make such directions and all necessary awards in the same way as if the procedure of the High Court as to joining one or more defendants or joining co-defendants or third parties was available to the parties and to him.

.2 Save that the Contractor or the Sub-Contractor may require the dispute or difference under this Sub-Contract to be referred to a different Arbitrator (to be appointed under this Sub-Contract) if either of them reasonably considers that the Arbitrator appointed to determine the related dispute is not appropriately qualified to determine the dispute or difference under this Sub-Contract.

.3 Articles 3.2.1 and 3.2.2 shall apply unless in the Tender, Schedule 1, item 10 the words 'Articles 5.1.4 and 5.1.5 apply' have been deleted.

Time for commencement of arbitration

Article 3.3 provides for arbitration not to commence until after practical completion or abandonment of the main contract works except on the matters listed below:

3.3 Such Arbitrator shall not without the written consent of the Contractor and Sub-Contractor enter on the arbitration until after the practical completion or abandonment of the Main Contract Works, except to arbitrate:

.1 whether a payment has been improperly withheld or is not in accordance with the Sub-Contract; or

.2 whether practical completion of the Sub-Contract Works shall be deemed to have taken place under clause 14.2; or

.3 in respect of a claim by the Contractor or counterclaim by the Sub-Contractor to which the provisions of clause 24 apply in which case the Arbitrator shall exercise the powers given to him in clause 24; or

.4 any matters in dispute under clause 4.3 in regard to reasonable objection by the Sub-Contractor or under clauses 11.2 and 11.3 as to extension of time.

Article 3.4 must be regarded as a safeguard because there are very few decisions of the architect which are final and binding:

3.4 In any such arbitration as is provided for in article 3 any decision of the Architect which is final and binding on the Contractor under the Main Contract shall also be and be deemed to be final and binding between and upon the Contractor and the Sub-Contractor.

Arbitration powers

Articles 3.5–3.7 generally endorse the powers of the arbitrator and the procedures which exist in the 1950 and 1979 Arbitration Acts which control the conduct of proceedings:

3.5 Subject to the provisions of clauses 4.6, 35.4.3, 36.5.3, 37.5 and clause 30 of the Main Contract Conditions the Arbitrator shall, without prejudice to the generality of his powers, have power to direct such measurements and/or valuations as may in his opinion be desirable in order to determine the rights of the parties and to ascertain and award any sum which ought to have been the subject of or included in any certificate and to open up, review and revise any certificate, opinion, decision, requirement or notice and to determine all matters in dispute which shall be submitted to him in the same manner as if no such certificate, opinion, decision, requirement or notice had been given.

3.6   The award of such Arbitrator shall be final and binding on the parties.

3.7   Whatever the nationality, residence or domicile of the Employer, the Contractor, the Sub-Contractor, and any sub-contractor or supplier or the Arbitrator, and wherever the Main Contract Works or Sub-Contract Works, or any parts thereof, are situated, the law of England shall be the proper law applicable to the Sub-Contract and in particular (but not so as to derogate from the generality of the foregoing) the provisions of the Arbitration Acts, 1950 (notwithstanding anything in S.34 thereof) to 1979 shall apply to any arbitration under article 3 wherever the same, or any part of it, shall be conducted.

Set off and adjudication under NSC/4

In the event of the contractor having a claim against the sub-contractor for one or more of a number of reasons the contractor may, understandably, feel disposed to deduct the sum he believes to be the value of his claim from any sum which may be due to the sub-contractor for work which he has carried out. Nominated Sub-contract/4, clause 23 makes provision for such event but clause 24 also provides a safeguard for the sub-contractor by providing for the appointment of an adjudicator to decide the course of action to be taken regarding any amount which the contractor may be disposed to deduct. The provisions of clause 24 arise largely because of abuse by contractors in the past of their ability to 'contra charge'. Sub-contractors were vulnerable and poorly placed to resist such deductions.

Contractor's right to set-off

Nominated sub-contract/4, clause 23.1 entitles the contractor to make deduction:

23.1   The Contractor shall be entitled to deduct from any money (including any Sub-Contractor's retention, notwithstanding the fiduciary obligation of the Contractor under clause 21.9.1) otherwise due under the Sub-Contract any amount agreed by the Sub-Contractor as due to the Contractor, or finally awarded in arbitration or litigation in favour of the Contractor, and which arises out of or under the Sub-Contract.

An important aspect of this clause is the words' . . . *agreed by the Sub-Contractor* as due to the Contractor or finally awarded . . . etc.' (my italics). The contractor is *not* entitled to deduct just any sum *he* considers to be appropriate. Clause 23.2 extends the contractor's powers to allow him to deduct amounts *not* agreed but it imposes three restrictions:

23.2   The Contractor shall be entitled to set off against any money (including any Sub-Contractor's retention, notwithstanding the fiduciary obligation of the Contractor under clause 21.9.1) otherwise due under the Sub-Contract the amount of any claim for loss and/or expense which has actually been incurred by the Contractor

211

by reason of any breach of, or failure to observe the provisions of, the Sub-Contract by the Sub-Contractor provided:

.1 that no set-off relating to any delay in completion shall be made unless, in accordance with clause 12, the certificate of the Architect referred to in clause 12.2 has been issued to the Contractor with a duplicate copy to the Sub-Contractor;

.2 the amount of such set-off has been quantified in detail and with reasonable accuracy by the Contractor;

.3 the Contractor has given to the Sub-Contractor notice in writing specifying his intention to set off the amount quantified in accordance with clause 23.2.2 and the grounds on which such set-off is claimed to be made. Such notice shall be given not less than 20 days before the money from which the amount or part thereof is to be set-off becomes due and payable to the Sub-Contractor; provided that such written notice shall not be binding in so far as the Contractor may amend it in preparing his pleadings for any arbitration pursuant to the notice of arbitration referred to in clause 24.1.1.1.

Clause 12, referred to in clause 23.2.1, above, is concerned with the 'failure of the sub-contractor to complete on time'. The sub-contractor's position in this respect is discussed in Chapter 13. An important feature of clause 23.2.1, above, is the issue of an architect's certificate.

Clauses 23.2.2 and 23.2.3 are restrictions which are concerned more with procedures than with the circumstances which give rise to the contractor's right to set-off.

NSC/4, clause 23.3 makes it clear that any agreements made by the contractor and sub-contractor regarding the amount set off is 'without prejudice' to subsequent negotiations and (perhaps more important) arbitration proceedings or litigation:

23.3 Any amount set off under the provisions of clause 23.2 is without prejudice to the rights of the Contractor or Sub-Contractor in any subsequent negotiations, arbitration proceedings or litigation to seek to vary the amount claimed and set-off by the Contractor under clause 23.2.

The concluding sub-clause of NSC/4, clause 23 excludes any *implied* terms of set-off:

23.4 The rights of the parties to the Sub-Contract in respect of set-off are fully set out in Sub-Contract NSC/4 and no other rights whatsoever shall be implied as terms of the Sub-Contract relating to set-off.

Contractor's claims not agreed by the sub-contractor

Appointment of adjudicator

NSC/4, clause 24 is concerned with the procedures for the appointment of a third party, the *adjudicator*, who will decide matters listed in clause 24.3 below. An adjudicator is *not* the arbitrator but a person called in to give a quick decision, on the basis of written evidence and usually without actually meeting the parties. Time is important to the contractor and sub-contractor in this respect.

Sub-Contractor's action

Clause 24.1 sets out the action which the sub-contractor should take if he disagrees with the set off:

24.1 .1   If the Sub-Contractor, at the date of the written notice of the Contractor issued under clause 23.2.3, disagrees the amount (or any part thereof) specified in that notice which the Contractor intends to set off, the Sub-Contractor may, within 14 days of receipt by him of such notice, send to the Contractor by registered post or recorded delivery a written statement setting out the reasons for such disagreement and particulars of any counterclaim against the Contractor arising out of the Sub-Contract to which the Sub-Contractor considers he is entitled, provided always that he shall have quantified such counterclaim in detail and with reasonable accuracy (which statement and counterclaim if any, shall not however be binding in so far as the Sub-Contractor may amend it in preparing his pleadings for any arbitration pursuant to the notice of arbitration referred to in clause 24.1.1.1) and shall at the same time:

  .1 .1   give notice of arbitration to the Contractor; and

  .1 .2   request action by the Adjudicator in accordance with the right given in clause 24.1.2 (and immediately inform the Contractor of such request) and send to the Adjudicator by registered post or recorded delivery a copy of the aforesaid statement and the written notice of the Contractor to which that statement relates and the aforesaid counterclaim (if any).

24.1 .2   Subject to the provisions of clause 24 and of clauses 21.3 and 23 the Sub-Contractor shall be entitled to request the Adjudicator named in the Tender, Schedule 2, item 6 to act as the Adjudicator to decide those matters referable to the Adjudicator under the provisions of clause 24. In the event of the above-named being unable or unwilling to act as the Adjudicator a person appointed by the above-named shall be the Adjudicator in his place. Provided that no person shall act as the Adjudicator who has any interest in the Sub-Contract or the Main Contract of which the Sub-Contract is part or in other contracts or sub-contracts in which the Contractor or the Sub-Contractor is engaged unless the Contractor, Sub-Contractor and the Adjudicator so interested otherwise agree in writing within a reasonable time of the Adjudicator's interest becoming apparent.

Clause 24.1 requires the sub-contractor to have facts and figures available before he acts. The fact that adjudication is frequently a prelude rather than an alternative to arbitration is apparent from clause 24.1.1.1 in which the sub-contractor is required to give notice of arbitration to the contractor.

Contractor's action

Clause 24.2 provides for the contractor's response to the sub-contractor's action:

24.2　Upon receipt of the aforesaid statement the Contractor may within 14 days from the date of such receipt send to the Adjudicator by registered post or recorded delivery a written statement with a copy to the Sub-Contractor setting out brief particulars of his defence to any counterclaim by the Sub-Contractor.

It appears from clause 24.2 that the contractor has fourteen days in which to *send* his statement to the adjudicator, i.e. time in transit for the statement should be added to the fourteen days. The contractor would, however, be wise to clarify this matter with the adjudicator to prevent the possibility of misunderstanding arising. The reason for this suggestion is obvious from clause 24.3.1 which indicates that when the stated time has elapsed the adjudicator makes his decision.

Adjudicator's action

Clause 24.3 describes the action to be taken by the adjudicator:

24.3　.1　Within 7 days of receipt of any written statement by the Contractor under clause 24.2 or on the expiry of the time limit to the Contractor referred to in clause 24.2 whichever is the earlier, the Adjudicator, without requiring any further statements than those submitted to him under clause 24.1 and where relevant clause 24.2 (save only such further written statements as may appear to the Adjudicator to be necessary to clarify or explain any ambiguity in the written statements of either the Contractor or the Sub-Contractor) and without hearing the Contractor or Sub-Contractor in person, shall, subject to clause 24.3.2, in his absolute discretion and, without giving reasons, decide, in respect of the amount notified by the Contractor under clause 23.2.3, whether the whole or any part of such amount shall be dealt with as follows:

24.3　.1　.1　shall be retained by the Contractor; or

　　　.1　.2　shall, pending arbitration, be deposited by the Contractor for security with the Trustee-Stakeholder named in the Tender, Schedule 2, item 6; or

　　　.1　.3　shall be paid by the Contractor to the Sub-Contractor; or

　　　.1　.4　any combination of the courses of action set out in clauses 24.3.1.1, 24.3.1.2 and 24.3.1.3.

　　　　　The Adjudicator's decision shall be binding upon the Contractor and the Sub-Contractor until the matters upon which he has given his decision have been settled by agreement or determined by an Arbitrator or the court.

24.3　.2　The Adjudicator shall reach such decision under clause 24.3.1 as he considers to be fair, reasonable and necessary in all the circumstances of the dispute as set out in the statements referred to in clauses 24.1 to .3, and such decision shall deal with the whole amount set off by the Contractor under clause 23.2

24.3 .3 The Adjudicator shall immediately notify in writing the Contractor and the Sub-Contractor of his decision under clause 24.3.1.

Clause 24.3 does not state the time in which the adjudicator should reach his decision but the clause conveys an impression of urgency which is emphasized by clause 24.3.3. He has three courses of action open to him when making his decision, four if he decides on the 'combination of the courses of action' set out in clause 24.3.1.4.

Implementation of adjudicator's decision

The two courses of action which involve the contractor making payments are set out in clause 24.4:

24.4 .1 Where any decision of the Adjudicator notified under clause 24.3.3 requires the Contractor to deposit an amount with the Trustee-Stakeholder, the Contractor shall thereupon pay such amount to the Trustee-Stakeholder to hold upon the terms hereinafter expressed provided that the Contractor shall not be obliged to pay a sum greater than the amount due from the Contractor under clause 21.3 in respect of which the Contractor has exercised the right of set-off referred to in clause 23.2.

24.4 .2 Where any decision of the Adjudicator notified under clause 24.3.3 requires the Contractor to pay an amount to the Sub-Contractor, such amount shall be paid by the Contractor immediately upon receipt of the decision of the Adjudicator but subject to the same proviso as set out in clause 24.4.1.

It appears that whilst the contractor is 'not obliged to pay a sum greater than the amount due to [him] . . .' there is nothing to prevent the adjudicator from deciding that a greater sum should be paid.

Trustee-stake holder's action

This is contained in clause 24.5:

24.5 .1 The Trustee-Stakeholder shall hold any sum received under the provisions of clauses 24.3 and 24.4 in trust for the Contractor and Sub-Contractor until such time as:

.1 .1 the Arbitrator appointed pursuant to the notice of arbitration given by the Sub-Contractor under clause 24.1.1.1; or

.1 .2 the Contractor and Sub-Contractor in a joint letter signed by each of them or on their behalf,

shall otherwise direct and shall, in either of the above cases, forthwith dispose of the said sums as may be directed by the Arbitrator, or failing any direction by the Arbitrator, as the Contractor and Sub-Contractor shall jointly determine. The Trustee-Skateholder shall deposit the sum received in a deposit account in the name of the Trustee-Stakeholder but shall add the interest to the sum deposited. The Trustee-Stakeholder shall be entitled to deduct his reasonable and proper charges from the sum deposited (including any interest added thereto). The Sub-Contractor shall notify the Trustee-Stakeholder of the name and address of the Adjudicator and Arbitrator referred to in clause 24.

24.5 .2 Where the Trustee-Stakeholder is a deposit-taking Bank then sums so received by it under the provisions of clauses 24.3 and 24.4 may, notwithstanding the trust imposed, be held by the Trustee-Stakeholder as an ordinary bank deposit to the credit of an account of the Bank as a Trustee-Stakeholder re the Contractor and Sub-Contractor referred to herein; and in respect of such deposit the Trustee-Stakeholder shall pay such usual interest which shall accrue to and form part of the deposit subject to the right of the Trustee-Stakeholder to deduct its reasonable and proper charges and any tax in respect of such interest from the sum deposited.

This action only becomes necessary if the adjudicator's decision is a course of action set out in either of clauses 24.3.1.2 or 24.3.1.4. When making his award the arbitrator will decide on the disposal of the sum deposited with the trustee-stakeholder together with the disposal of any interest which may have accrued.

Arbitrator's power to act

Clause 24.6 states the arbitrator's power to vary or cancel the decision of the adjudicator if either party applies to him to do so:

24.6 The Arbitrator appointed pursuant to the notice of arbitration given under clause 24.1.1.1 may in his absolute discretion at any time before his final award on the application of either party vary or cancel the decision of the Adjudicator given under clause 24.3 if it appears just and reasonable to him so to do.

The arbitrator would make an interim award if he considered it reasonable for him to do so. This, again, indicates that the adjudicator's powers are subordinate to those of the arbitrator and that he is, in effect, conducting a holding action of a temporary nature.

Further claims and counter claims

The action described above is by no means a 'once and for all time' exercise. Should the sub-contractor have further cause to disagree with set-off action which the contractor proposes taking he may, again, take the action described in clause 24.1. Clause 24.7 reads:

24.7 Any action taken by the Contractor under clause 23.2 and by the Sub-Contractor in respect of any counterclaim under clause 24.1.1 is without prejudice to similar action by the Contractor or Sub-Contractor as the case may be if and when further sums become due to the Sub-Contractor.

Adjudicator's fees

It is the sub-contractor who initiates action involving the adjudicator and it is the sub-contractor who is responsible for paying the adjudicator's fee until the arbitrator makes his final award. Clause 24.8 reads:

24.8   The fee of the Adjudicator shall be paid by the Sub-Contractor but the Arbitrator appointed pursuant to the notice of arbitration under clause 24.1.1.1 shall in his final award settle the responsibility of the Contractor or Sub-Contractor or both for payment of the fee or any part thereof and where relevant for the charges of the Trustee-Stakeholder or any part thereof.

Adjudicators and trustee-stakeholders' nomination

Schedule 2 of NSC/1, item 6, provides space for the insertion of the names of an adjudicator and a trustee-stakeholder. These names should be agreed between the parties and inserted when the contractor and sub-contractor enter into their contract. The NFBTE and the various specialist contractors' associations hold lists of persons who are prepared to act as adjudicators. Trustee-stakeholders are normally deposit-taking banks. The trustee-stakeholder should be informed of any conditions imposed by the conditions of the sub-contract.

Arbitration and set-off for contracts under DOM/1

Arbitration

Arbitration is the means of settling disputes which is provided in article 3 of DOM/1. DOM/1 and NSC/4 have essentially the same intentions in their respective article 3s but article 3.2 in DOM/1 makes no mention of the other nominated sub-contract documents (NSC/2, 2a, 4 or 4a) nor does it provide for the 'different Arbitrator' to which there is reference in article 3.2.2 of NSC/4.

Set off and adjudication

Set off and adjudication under contract DOM/1 and NSC/4 are essentially the same but NSC/4, clause 23.2.1 does not appear in DOM/1. This is because the architect has no cause to certify delay on domestic sub-contractor's work.

Arbitration for contracts not under sub-contracts NSC/4 or DOM/1

Clause 3.7, above, refers to the Arbitration Acts of 1950 and 1979. If the parties to *any* contract agree, whether or not they have a formal contract, they may submit a dispute to arbitration under the Arbitration Acts. In this event the provisions of the Acts would be applied. This would have much the same effect as a submission made under NSC/4, article 3.

Chapter 15

Interim certificates and payments, completion and final payment

Payments on account, as the works proceed, have been a regular feature of building projects since earliest recorded history. Unlike the works of most other producers, the products of the numerous specialist craftsmen – contractors who contribute to building works are *fixed to land* which is owned by others, frequently the employer – the person/s commissioning the building works, so for most practical purposes the specialist contractor's product becomes the employer's property (real estate) as soon as it is fixed in position. It is only practical that it should be paid for as and when it is fixed.

Furthermore, building works are frequently costly and it is therefore beyond the means of many specialist contractors to *complete* their works before they receive payment.

Most standard forms of building contract provide for payments on account but sub-contractors should bear in mind the warnings given in Chapter 6 that payments on account should not be taken for granted on projects where a standard form is not in use.

Interim payments using the standard forms

Payments to sub-contractors are made 'from time to time' (SF 80, cl. 30.1.1.1), that is to say *interim* payments, and, when the works are complete and satisfactory, a final payment is made.

The general arrangements and procedures for payments are contained in the main contract conditions, SF 80, clause 30. This clause is primarily concerned with certificates and payments to the *contractor* but as there are numerous references to this clause in NSC/4 it must provide the starting point for the sub-contractor.

Architect issues interim certificates

The obligation on the architect to issue interim certificates is contained in SF 80, clause 30.1.1.1. which indicates the obligation ('The Architect *shall* . . .') and the entitlement of the contractor with regard to the timing of the payment:

30.1 .1 .1 The Architect shall from time to time as provided in clause 30 issue Interim Certificates stating the amount due to the Contractor from the Employer and the Contractor shall be entitled to payment therefor within 14 days from the date of issue of each Interim Certificate.

Employer obliged to honour architect's certificate

The employer is bound by the architect's certificate to make payment to the contractor but the employer has various rights of deduction which are set out elsewhere in the conditions. For most practical purposes, however, the employer is bound by the certificate and when he exercises his rights of deduction he is required to inform the contractor in writing of the reasons for such deductions (SF 80, cl.30.1.1.3).

Interim valuations made by quantity surveyor

SF 80, clause 30.1.2 states the procedure succinctly:

30.1 .2 Interim valuations shall be made by the Quantity Surveyor whenever the Architect considers them to be necessary for the purpose of ascertaining the amount to be stated as due in an Interim Certificate.

The contractor normally cooperates with the quantity surveyor in preparing the valuations.

Period of interim certificates

SF 80, clause 30.1.3 states:

30.1 .3 Interim Certificates shall be issued at the Period of Interim Certificates specified in the Appendix up to and including the end of the period during which the Certificate of Practical Completion is issued. Thereafter Interim Certificates shall be issued as and when further amounts are ascertained as payable to the Contractor from the Employer and after the expiration of the Defects Liability Period named in the Appendix or upon the issue of

The period '. . . specified in the Appendix . . .', if none is stated, is one month. It is most unusual for the period specified to be other than 'one month'.

Ascertainment of the amounts due in interim certificates

The amount of an interim certificate is the 'gross valuation' which includes:

(a) the value of the work properly executed by the contractor (SF 80, cl. 30.2.1.1);
(b) the value of the materials and goods delivered to the site (SF 80, cl. 30.2.1.2);
(c) the value of materials and goods off-site but set apart for the works (SF 80, cl. 30.2.1.3);

(d) amounts due to nominated sub-contractors (SF 80, cl. 30.2.1.4);

(e) the contractor's profit on (d) above (SF 80, cl. 30.2.1.5).

Items (a)–(e), above, are all subject to 'retention' to which reference is made below.

In addition to items (a)–(e), the gross valuation includes: items which are *not* subject to retention, two of which are of interest to sub-contractors. These are:

(f) final payments made to sub-contractors who completed their work in advance of completion of the contractor's works (SF 80, cl. 30.2.2.3);

(g) the amounts referred to in clause 21.4.2 of NSC/4 to which reference is made below (SF 80, cl. 30.2.2.5).

The 1980 edition of the Standard Form of Building Contract, clause 30.2.3 states that the $2\frac{1}{2}$ per cent discount for prompt payment which the contractor is entitled to deduct from amounts due to nominated sub-contractors is not subject to retention.

Items (a)–(g) above should be sufficiently informative for the sub-contractor's purposes but they are merely summaries of the clauses contained in SF 80 and the sub-contractor should study the actual conditions of contract for more detail.

Retention

The intention of the conditions of contracts is that whilst the works are in progress the employer retains a percentage of the total amount due to the contractor. The percentage is usually 5 per cent but a lower rate may be agreed. A footnote to SF 80, clause 30.4.1.1 recommends that 'where the employer at the tender stage estimates the Contract Sum to be £500,000 or over, the Retention Percentage should not be more than 3 per cent'. Whatever sum is agreed between the employer and the contractor should be specified in the appendix to SF 80. The sub-contractor is entitled to view the appendix to the main contract to satisfy himself with regard to the matters which the employer and the contractor have agreed. This is an opportunity which the sub-contractor should not overlook.

The retention percentage may be deducted from the amount due which relates to work which *has not* reached practical completion and amounts in respect of materials and goods (SF 80, cl. 30.4.1.2).

Half the retention percentage may be deducted from so much of the total amount which *has* reached practical completion (SF 80, cl. 30.4.1.3).

A footnote to clause 30.4.1.2 explains the operation of deductions of the retention percentage: '... the Contractor will have released to him by the Employer upon payment of the next Interim Certificate after Practical Completion of the whole or part of the Works – approximately one-half of the Retention on the whole or the appropriate part; and upon payment of the next Interim Certificate after the expiration of Defect Liability Period named in the Appendix, or after the issue of the Certificate of Completion of Making Good Defects, whichever is the later, the balance of the Retention on the whole or the appropriate part. When Retention is so included in Interim Certificates it becomes a 'sum due' to the Contractor and therefore subject to the rights of the Employer to deduct therefrom in accordance with the rights of the Employer so to deduct as set out in the Conditions.'

The employer holds the retention in trust but he is not obliged to invest it (SF 80, cl. 30.5.1).

The architect is required to state in his interim certificates the retention which is to be deducted from amounts due to the contractor and the nominated sub-contractors and to issue such a statement to the employer, to the contractor and to each nominated sub-contractor (SF 80, cl. 30.5.2).

Interim payment to sub-contractor

So much for the procedures regarding interim valuations, certificates and payments in SF 80 generally. Clause 21 in both NSC/4 and DOM/1 relate to payment of the sub-contractor. The clauses contained in NSC/4 have been used below to explain the conditions but as these vary in some respects from DOM/1, domestic sub-contractors should refer to the conditions contained in the latter form. NSC/4, clause 21.2.1 gives the sub-contractor the right to apply for payment:

21.2 .1 Notwithstanding the requirement that the Architect shall issue Interim Certificates under clause 30 of the Main Contract Conditions, the Contractor shall, if so requested by the Sub-Contractor, make application to the Architect as to the matters referred to in clauses 30.2.1.4, 30.2.2.5 and 30.2.3.2 of the Main Contract Conditions.

The clauses listed in the last line, above, refer to amounts which are not subject to retention – see under heading 'Retention', above. The contractor is obliged to pass on to the architect any application made by the sub-contractor:

21.2 .2 The Contractor shall include in or annex to any application under clause 21.2.1 any written representations of the Sub-Contractor which the Sub-Contractor wishes the Architect to consider including those referred to in clause 37.3.2.

Clause 37.3.2, referred to in the last line, above, is concerned with *fluctuations*.

Payment for off-site goods

NSC/4, clause 21.2.3 refers to payment for off-site goods and materials:

21.2 .3 The Sub-Contractor shall observe any relevant conditions in clause 30.3 of the Main Contract Conditions before the Architect is empowered to include the value of any off-site materials or goods in Interim Certificates.

221

'Clause 30.3 of the Main Contract Conditions' states the conditions which sub-contractors must observe. These are:

30.3 .1 the materials are intended for incorporation in the Works.

30.3 .2 nothing remains to be done to the materials to complete the same up to the point of their incorporation in the Works.

30.3 .3 the materials have been and are set apart at the premises where they have been manufactured or assembled or are stored, and have been clearly and visibly marked, individually or in sets, either by letters or figures or by reference to a pre-determined code, so as to identify:

.3 .1 the Employer, where they are stored on the premises of the Contractor, and in any other case the person to whose order they are held; and

.3 .2 their destination as the Works;

30.3 .4 where the materials were ordered from a supplier by the Contractor or by any Sub-Contractor, the contract for their supply is in writing and expressly provides that the property therein shall pass unconditionally to the Contractor or the Sub-Contractor (as the case may be) not later than the happening of the events set out in clauses 30.3.2 and 30.3.3.

30.3 .5 where the materials were ordered from a supplier by any sub-contractor, the relevant sub-contract between the Contractor and the Sub-Contractor is in writing and expressly provides that on the property in the materials passing to the sub-contractor the same shall immediately thereon pass to the Contractor.

30.3 .6 where the materials were manufactured or assembled by any Sub-Contractor, the Sub-Contract is in writing and expressly provides that the property in the materials shall pass unconditionally to the Contractor not later than the happening of the events set out in clauses 30.3.2 and 30.3.3.

30.3 .7 the materials are in accordance with this Contract.

30.3 .8 the Contractor provides the Architect with reasonable proof that the property in the materials is in him and that the appropriate conditions set out in clauses 30.3.1 to .7 have been complied with.

30.3 .9 the Contractor provides the Architect with reasonable proof that the materials are insured against loss or damage for their full value under a policy of insurance protecting the interests of the Employer and the Contractor in respect of the Clause 22 Perils, during the period commencing with the transfer of property in the materials to the Contractor until they are delivered to, or adjacent to, the Works.

These conditions are intended to safeguard the employer's interests.

Most quantity surveyors will wish to visit the place 'off-site' where the goods are being stored to satisfy themselves that the conditions are being complied with before they will include an amount in the valuation for the goods in question.

Payment of nominated sub-contractors

The 1980 edition of the Standard clause 35.13 describes the action which the employer, the architect, and the contractor should take in respect of payment of nominated sub-contractors. SF 80, the main contract, is between employer and contractor and clause 35.13 will not be discussed in detail except where it is referred to in NSC/4. Nevertheless, the sub-contractor should be aware of the general contents of the clause which are for the architect to direct the contractor as to the amount of interim and final payment and for the contractor to provide the architect with reasonable proof of discharge of each interim payment. In the event of the contractor not being able to provide reasonable proof, clause 35.13.5 empowers the employer to pay the sub-contractor direct. This is a condition which the sub-contractor should take care to remember.

NSC/4 clause 21.3.1.1 makes a clear statement of the action to be taken for payment of the sub-contractor by the contractor:

21.3 .1 .1 Within 17 days of the date of issue of an Interim Certificate (including the Interim Certificate referred to in clause 35.17 or clause 30.7 of the Main Contract Conditions) the Contractor shall notify to the Sub-Contractor the amount included in the amount stated as due therein in respect of the Sub-Contract Works and shall duly discharge his obligation to pay the Sub-Contractor such amount less only a cash discount of $2\frac{1}{2}$ per cent if discharge is effected within the said 17 days. Immediately upon discharge by the Contractor as aforesaid the Sub-Contractor shall supply the Contractor with written proof of such discharge so as to enable the Contractor to provide the Architect with the 'reasonable proof' referred to in clause 35.13.3 of the Main Contract Conditions. [k]

Footnote [k] above, draws attention to clause 23 (contractor's right to set off, which is discussed in Chapter 14) and to clauses 20A or 20B (tax deduction scheme) which may be relevant to due discharge by the contractor.

The sub-contractor should take care to provide the 'written proof' as quickly as possible to prevent delay in certification of the subsequent payment.

The employer's right to deduct monies 'in respect of some act or default by the sub-contractor' is set out in NSC/4, clause 21.3.1.2.

.1 .2 Where the Employer has exercised any right under the Main Contract to deduct from monies due to the Contractor and such deduction is in respect of some act or default of the Sub-Contractor, his servants or agents the amount of such deduction may be deducted by the Contractor from any monies due or to become due under the Sub-Contract or may be recoverable by the Contractor from the Sub-Contractor as debt.

The importance, from the sub-contractor's standpoint, of the contract using the basic method of nomination, as described in Chapter 5, can be appreciated when reading NSC/4, clause 21.3.2.1.

21.3 .2 .1 The Contractor shall only be under an obligation duly to discharge any amount certified in an Interim Certificate issued under clause 35.17 of the Main Contract Conditions provided the Sub-Contractor shall have entered into Agreement NSC/2 including clause 5 of that Agreement unamended in any way and such Agreement is in full force and effect.

 .2 .2 Where the Contractor is under an obligation duly to discharge any amount certified in an Interim Certificate issued under clause 35.17 of the Main Contract Conditions the Sub-Contractor upon such discharge hereby agrees to indemnify the Contractor in respect of any omission, fault or defect in the Sub-Contract Works caused by the Sub-Contractor, his servants or agents for which the Contractor may at any time become liable to the Employer but subject always to the terms of clause 35.19.1 of the Main Contract Conditions.

Clause 35.17, referred to in the above clause, is concerned with the contractor making final payment to the sub-contractor in advance of the contractor's own contract completion.

Ascertainment of amounts due in sub-contractor's interim certificates

The amount of an interim certificate to the sub-contractor is set out in NSC/4, clause 21.4:

21.4 Subject to any agreement between the Sub-Contractor, the Contractor and the Architect as to stage payments, the amount of an interim payment to the Sub-Contractor which is included in the amount stated as due in an Interim Certificate and to which the provisions of clause 35.13 of the Main Contract Conditions apply shall be the gross valuation as referred to in clause 21.4 less an amount equal to any amount which may be deducted and retained by the Employer as provided in clause 30.2 of the Main Contract Conditions (referred to in the Main Contract Conditions as 'the Retention') in respect of the Sub-Contract Works; and
the total amount in respect of the Sub-Contract Works included in the total amount stated as due in Interim Certificates previously issued under the Main Contract Conditions.

Passing reference was made, above, to the composition of the contractor's gross valuation: NSC/4, clause 21.4 details the composition of the gross valuation for the sub-contractor:

The gross valuation shall be the total of the amounts referred to in clauses 21.4.1 and 21.4.2 less the total amount referred to in clause 21.4.3 as applied up to and including a date not more than 7 days before the date of the Interim Certificate:

21.4 .1 .1 the total value of the sub-contract work properly executed by the Sub-Contractor, including any work so executed to which clause 16.1 refers, together with, where applicable, any adjustment of that value under clause 37.

 .1 .2 the total value of the materials and goods delivered to or adjacent to the Works for incorporation therein by the Sub-Contractor but not so incorporated provided that the value of such materials and goods shall

224

only be included as and from such times as they are reasonably, properly and not prematurely so delivered and are adequately protected against weather and other casualties.

.1 .3 the total value of any materials or goods other than those to which clause 21.4.1.2 refers where the Architect in the exercise of his discretion under clause 30.3 of the Main Contract Conditions has decided that such total value shall be included in the amount stated as due in an Interim Certificate.

.2 .1 any amount to be included in Interim Certificates in accordance with clause 3 as a result of payments made or costs incurred by the Sub-Contractor under clause 6 or 7 of the Main Contract Conditions as referred to in clause 5.1.1, and under clause 14.4.

.2 .2 any amount ascertained as a result of the application of clause 13.1.

.2 .3 any amount payable to the Sub-Contractor under clauses 35 or 36, whichever is applicable.

.2 .4 an amount equal to one thirty-ninth of the amounts referred to in clauses 21.4.2.1, 21.4.2.2 and 21.4.2.3.

21.4 .3 Any amount allowable by the Sub-Contractor to the Contractor under clause 35 or 36, whichever is applicable together with an amount equal to one thirty-ninth of that amount.

Clause 21.4.1.1 above makes provision for the inclusion of variations (see Ch. 12).

The clauses referred to in clause 21.4.2 above are:

Clause 3 additions to or deductions from the Sub-Contract Sum arising from fees or charges in connection with statutory obligations and all provisions of the main contract,
Clause 13.1 direct loss and/or expense
Clauses 35 or 36 fluctuations

Retention

The rules for ascertainment of retention are those referred to in SF 80, clause 30.2 which are discussed above (NSC/4, cl. 21.5 and 21.6).

Disputes as to certificates

In the event of the sub-contractor feeling aggrieved he has recourse to arbitration:

21.7 If the Sub-Contractor shall feel aggrieved by any amount certified by the Architect or by his failure to certify, then subject to the Sub-Contractor giving to the Contractor such indemnity and security as the Contractor shall reasonably require, the Contractor shall allow the Sub-Contractor to use the Contractor's name and if necessary will join with the Sub-Contractor in arbitration proceedings at the instigation of the Sub-Contractor in respect of the said matters complained of by the Sub-Contractor.

He also has the right to suspend execution of the sub-contract works if the contractor fails to discharge his obligation to make payment (NSC/4, cl. 21.8),

21.8 .1 If:

    .1   .1   subject to clause 23 the Contractor shall fail to discharge his obligation to make any payment to the Sub-Contractor as hereinbefore provided; and

    .1   .2   the Employer has

        either for any reason not operated the provisions of clause 35.13.5 of the Main Contract Conditions,

        or has operated those provisions but for any reason has not paid the Sub-Contractor direct the whole amount which the Contractor has failed to discharge,

        within 35 days from the date of issue of the Interim Certificate in respect of which the Contractor has so failed to make proper discharge of his obligation in regard to payment of the Sub-Contractor,

    then provided the Sub-Contractor shall have given 14 days notice in writing to the Contractor and the Employer of his intention to suspend the further execution of the Sub-Contract Works, the Sub-Contractor may (but without prejudice to any other right or remedy) suspend the further execution of the Sub-Contract Works until such discharge or until such direct payment is made whichever first occurs.

    .2   Such period of suspension shall not be deemed a delay for which the Sub-Contractor is liable under the Sub-Contract. The Contractor shall be liable to the Sub-Contractor for any loss, damage or expense caused to the Sub-Contractor by any suspension of the Sub-Contract Works under the provisions of clause 21.8.1. The right of the Sub-Contractor under clause 21.8.1 shall not be exercised unreasonably or vexatiously.

The sub-contractor should take care to follow the procedures.

The crucial periods of time contained in the clause are

(a)   that thirty-five days have elapsed since the date of the issue of the architect's interim certificate; and

(b)   that the sub-contractor must give the contractor fourteen days notice of his intention to suspend work.

There is no reason why the sub-contractor should not give notice as soon as the seventeen days referred to in NSC/4, clause 21.3.1.1 have passed. In this way he will be in a position to suspend as soon as the thirty-five days have elapsed.

The contractor holds the sub-contractor's retention in trust in much the same way as the employer holds the main contract retention in trust. NSC/4, clause 21.9 describes the contractor's fiduciary interest in the sub-contractor's retention and the action which the contractor shall take in respect of unpaid retention money:

21.9 .1 The Contractor's interest in the Sub-Contractor's retention (as identified in the statement issued under clause 30.5.2 of the Main Contract Conditions and referred to in clause 21.6) is fiduciary as trustee for the Sub-Contractor (but without obligation to invest) and if the Contractor attempts or purports to mortgage or otherwise charge such interest or his interest in the whole of the amount retained as aforesaid (otherwise than by floating charge if the Contractor is a limited company) the Contractor shall thereupon immediately set aside in a separate bank account and become a trustee for the Sub-Contractor of a sum equivalent to the Sub-Contractor's retention as identified in the aforesaid statement; provided that upon payment of the same to the Sub-Contractor the amount due to the Sub-Contractor upon final payment under the Sub-Contract shall be reduced accordingly by the amount so paid.

21.9 .2 If any of the Sub-Contractor's retention is withheld by the Contractor after the period within which such retention should be discharged by the Contractor, the Contractor shall immediately upon the expiry of the aforesaid period place any such unpaid retention money in a separate trust account so identified as to make clear that the Contractor is the trustee for the Sub-Contractor of all such undischarged retention.

NSC/4, clause 21.9.2 recognizes that there may be reasons for the contractor *not* releasing the sub-contractor's retention as soon as he receives it from the employer but at the same time safeguards the sub-contractor's interests to some extent, at least.

Practical completion of sub-contract works – liability for defects

Sub-contractors occasionally have rather a raw deal regarding 'completion' of their works. They may complete their works months, even years, ahead of the contractor but they frequently have to wait an inordinate length of time before they obtain release of their retention monies. This is an occasion when a nominated sub-contractor is significantly less vulnerable than a domestic sub-contractor. Let us look at both NSC/4 and DOM/1 in this respect.

The sub-contractor undertakes, in article 1 of both the sub-contract forms, to 'carry out and *complete* the sub-contract works' (my italics) and it is therefore important that he appreciates the requirements of his contract with regard to practical completion which are contained in clause 14 of both the above-mentioned contracts. Dealing first with the conditions contained in NSC/4, clause 14.1 reads:

14.1 If the Sub-Contractor notifies the Contractor in writing of the date when in the opinion of the Sub-Contractor the Sub-Contract Works will have reached practical completion, the Contractor shall immediately pass to the Architect any such notification together with any observations thereon by the Contractor (a copy of which observations must immediately be sent by the Contractor to the Sub-Contractor).

It can be seen from the above clause that the onus is upon the sub-contractor to notify the contractor who in turn must notify the architect at which point it rests with the architect to name the day:

14.2   Practical completion of the Sub-Contract Works shall be deemed to have taken place on the day named in the certificate of practical completion of the Sub-Contract Works issued by the Architect under clause 35.16 of the Main Contract Conditions or as provided in clause 18.1.2 of the Main Contract Conditions.

Clause 35.16 of SF 80, mentioned above, requires the architect to issue the certificate when in his opinion practical completion of the sub-contract works is achieved and for him to send a duplicate copy of his certificate to the sub-contractor.

Clause 18.1.2 of SF 80, the other main contract clause mentioned above, is important because it signals the commencement of the defects liability period.

Sub-contractor's liability for defects under NSC/4

When reading clause 14.3 the sub-contractor should note the exclusions in the penultimate and last line.

14.3   Subject to clause 18 of the Main Contract Conditions but without prejudice to the obligation of the Sub-Contractor to accept a similar liability to any liability of the Contractor under the Main Contract to remedy defects in the Sub-Contract Works, the Sub-Contractor shall be liable to make good at his own cost and in accordance with any instruction of the Architect or direction of the Contractor all defects, shrinkages and other faults in the Sub-Contract works or in any part thereof considered necessary by reason of such defects, shrinkages or other faults due to materials or workmanship not in accordance with the Sub-Contract or due to frost occurring before the date of practical completion of the Sub-Contract Works.

The above wording is similar in many respects to that contained in the main contractor's contract, clause 17.2, which requires the architect to 'deliver to the Contractor as an instruction . . . not later than 14 days after the expiration of the Defects Liability Period' a schedule of defects which the contractor is to make good within a reasonable time at his own cost.

There are, however, occasions when the architect considers that the contractor should not pay the whole of the cost of making good and this benefit is extended to the sub-contractor in clause 14.4:

14.4   Where the Contractor is liable under the Main Contract to make good defects, shrinkages or other faults but the Architect instructs that it shall not be entirely at his own cost the Contractor shall grant a corresponding benefit to the Sub-Contractor to the extent that such benefit is relevant and applicable to the Sub-Contract Works.

When drafting conditions of contract nothing must be left unsaid but it is to be hoped that the sub-contractor would attend to clause 14.5 as a matter of good workmanship:

14.5 The Sub-Contractor upon practical completion of the Sub-Contract Works shall properly clear up and leave the Sub-Contract Works, and all areas made available to him for the purpose of executing those Works and, so far as used by him for that purpose, clean and tidy to the reasonable satisfaction of the Contractor.

The clauses regarding practical completion for domestic sub-contractors do not involve the architect as closely as those in NSC/4. DOM/1, clause 14.1 reads:

14.1 The Sub-Contractor shall notify the Contractor in writing of the date when in his opinion the Sub-Contract Works are practically completed and if not dissented from in writing by the Contractor within 14 days of receipt of the Sub-Contractor's written notice practical completion for all the purposes of the Sub-Contract shall be deemed to have taken place on the date so notified. Any written notice of dissent shall set out the Contractor's reasons for such dissent.

It will be noted that the contractor is not required to notify the architect and in clause 14.2 the sub-contractor's date for practical completion is linked to the contractor's date for practical completion:

14.2 When the Contractor gives written notice of dissent under clause 14.1 practical completion for all the purposes of this Sub-Contract will be deemed to have taken place on such date as may be agreed. Failing such agreement practical completion will be deemed to have taken place on the date of Practical Completion of the Works as certified by the Architect under clause 17.1 of the Main Contract Conditions.

Specialist sub-contractors accustomed to acting as *nominated* sub-contractors should note carefully the above condition because it may mean that they will have to wait longer for release of retention money than they would anticipate.

Sub-contractor's liability for defects under DOM/1

Clause 14.3 reads:

14.3 Subject to clause 18 of the Main Contract Conditions but without prejudice to the obligation of the Sub-Contractor to accept a similar liability to any liability of the Contractor under the Main Contract to remedy defects in the Sub-Contract Works, the Sub-Contractor shall be liable to make good at his own cost and in accordance with any direction of the Contractor all defects, shrinkages and other faults in the Sub-Contract Works or in any part thereof due to materials or workmanship not in accordance with the Sub-Contract or due to frost occurring before the date of practical completion of the Sub-Contract Works.

The above clause does not differ greatly from the similar clause in NSC/4 but the contractor directs the making good to be done and the architect is not mentioned.

229

The sub-contractor does, however, obtain the benefit of the architect's instructions as to the cost of making good defects. Domestic sub-contract/1, clause 14.4 reads:

14.4 Where the Contractor is liable under the Main Contract to make good defects, shrinkages or other faults but the Architect instructs that it shall not be entirely at his own cost the Contractor shall grant a corresponding benefit to the Sub-Contractor to the extent that such benefit is relevant and applicable to the Sub-Contract Works.

Final certificate and ascertained final sub-contract sum

The price for the sub-contract works may be expressed as either the *sub-contract sum* or the *tender sum* subject to complete remeasurement. The valuation of variations for both sums is discussed in Chapter 12. NSC/4, clause 21.10 provides for final adjustment of the *sub-contract sum*. NSC/4, clause 21.10.2 deals with preparation of a statement of the final valuation of variations (concluding the action taken in cl. 15), which the quantity surveyor is required to undertake within the 'Period of Final Measurement and Valuation' which 'if none stated is 6 months from the day named in the Certificate of Practical Completion of the Works' (appendix to SF 80). NSC/4, clause 21.10.1 reads:

21.10 .1 .1 Where clause 15.1 applies, either before or within a reasonable time after practical completion of the Sub-Contract Works the Sub-Contractor shall send to the Contractor or, if so instructed by him, to the Architect or the Quantity Surveyor, all documents necessary for the purpose of the adjustment of the Sub-Contract Sum.

.1 .2 Subject to compliance by the Sub-Contractor with the requirements of clause 21.10.1.1 a statement of the final valuation of all Variations under clause 16 will be prepared by the Quantity Surveyor either within the Period of Final Measurement and Valuation stated in the Appendix to the Main Contract Conditions (Tender, Schedule 1, item 10) or at such other time as is necessary to enable the provisions of clause 35.17 of the Main Contract Conditions to be operated and the Architect will send a copy to the Contractor and the Sub-Contractor.

The sub-contractor should take care to provide 'all documents necessary . . .' (see above) because his failure to do so may delay the final valuation. There is considerable evidence to suggest that sub-contractors are a major cause of delay in this respect. NSC/4, clause 21.10.2 contains a list of items to be included in adjustment of the sub-contract sum:

21.10 .2 The Sub-Contract Sum shall be adjusted as follows.
There shall be deducted:

.2 .1 all provisional sums and the value of all work described as provisional included in the Sub-Contract Documents.

.2 .2   the amount of the valuation under clause 16.3.2 of items omitted in accordance with a Variation required by an instruction of the Architect or subsequently sanctioned by him in writing together with the amount of any other work included in the Sub-Contract Documents as referred to in clause 16.3.5 which is to be valued under clause 16.3.

.2 .3   any amount allowed to the Contractor under clause 35, 36 or 37, whichever is applicable, together with an amount equal to one thirty-ninth of that amount.

.2 .4   any other amount which is required by the Sub-Contract Documents to be deducted from the Sub-Contract Sum.

There shall be added:

.2 .5   any amount paid or payable by the Contractor to the Sub-Contractor as a result of payments made or costs incurred by the Sub-Contractor under clause 6 or 7 of the Main Contract Conditions as referred to in clause 5.1.1 and under clause 14.1.

.2 .6   the amount of the valuation under clause 16.3 of any Variation including the valuation of other work, as referred to in clause 16.3.5, under clause 16.3 other than the amount of the valuation of any omission under clause 16.3.2.

.2 .7   the amount of the valuation of work executed by, or the amount of any disbursements made by, the Sub-Contractor in accordance with the instructions of the Architect as to the expenditure of provisional sums included in the Sub-Contract Documents, and of all work described as provisional included in the Sub-Contract Documents.

.2 .8   any amount ascertained as a result of the application of clause 13.1.

.2 .9   any amount paid or payable to the Sub-Contractor under clause 35, 36 or 37, whichever is applicable.

.2 .10 any other amount which is required by the Sub-Contract to be added to the Sub-Contract Sum.

.2 .11 an amount equal to one thirty-ninth of the amounts referred to in clauses 21.10.2.5, 21.10.2.8 and 21.10.2.9.

21.10 .3 Before the Architect certifies final payment for the Sub-Contract Works under clause 35.17 or 30.7 of the Main Contract Conditions the Sub-Contractor shall be supplied through the Contractor with a copy of the computation of the Sub-Contract Sum adjusted in accordance with clause 21.10.2.

The above clause provides a useful check-list. A typical format which might be used for the final statement is shown in Table 15.1.

NSC/4, clause 21.11 is concerned with *computation* of the ascertained final sub-contract sum. Clause 15.2, mentioned in the first line of clause 21.11.1 below, refers to 'complete re-measurement' sub-contracts:

21.11 .1   Where clause 15.2 applies, either before or within a reasonable time after practical completion of the Sub-Contract Works the Sub-Contractor shall send to the Contractor or, if so instructed by him, to the Architect or the

231

Quantity Surveyor, all documents necessary for the purpose of computing the Ascertained Final Sub-Contract Sum.

21.11 .2 The Ascertained Final Sub-Contract Sum shall be the aggregate of the following:

.2 .1 any amount paid or payable by the Contractor to the Sub-Contractor as a result of payments made or costs incurred by the Sub-Contractor under clause 6 or 7 of the Main Contract Conditions as referred to in clause5.1.1 and under clause 14.1.

.2 .2 the amount of the Valuation under clause 17.

.2 .3 any amount ascertained as a result of the application of clause 13.1

.2 .4 any amount paid or payable to or allowed or allowable by the Sub-Contractor under clause 35, 36 or 37, whichever is applicable.

.2 .5 any other amount which is required to be included or taken into account in computing the Ascertained Final Sub-Contract Sum.

.2 .6 an amount equal to one thirty-ninth of the amounts referred to in clauses 21.11.2.1, 21.11.2.3 and 21.11.2.4, so far as clause 21.11.2.4 refers to clauses 35 and 36.

21.11 .3 The Sub-Contractor shall be supplied with a copy of the computation of the Ascertained Final Sub-Contract Sum before the Architect certifies final payment for the Sub-Contract Works under clause 35.17 or 30.7 of the Main Contract Conditions.

Table 15.1 Final Account format

The final account general summary will probably read:

		Omissions (£)	Additions (£)
Item 1	Sub-contract sum	–	75,600.00
Item 2	Variations: as item 2 summary	2,100.00	7,250.00
Item 3	Expenditure of prime cost sums: as item 3 summary	3,500.00	4,100.00
Item 4	Price adjustment formula: as item 4 summary	–	13,760.00
			100,710.00
		5,600.00	5,600.00
			£95,110.00

There are two surprising points in connection with clause 21.11. The first is that the clause makes no reference to responsibility for preparation of the computation being with the quantity surveyor, as does clause 21.10.1.2.

The second is that the proviso which concludes clause 21.11.2.6 which refers to fluctuations does not appear in clause 21.10.2.11. This would appear to be a deliberate difference between the clauses because the proviso does not conclude either of the clauses in the original text of the JCT *Guide* but it is listed in the corrigenda. There are notable differences between the concluding sub-clauses of clauses 21.10 and 21.11 in so far as the supply of the architect's certificate is concerned.

Sub-contracts using DOM/1

DOM/1, clause 21 provides for *payment of sub-contractor*. The clause follows a similar, but by no means identical, pattern to that followed by NSC/4. clause 21. Whilst many of the comments made above provide guidance for the sub-contractor employed as a *domestic* sub-contractor, (and cl. 21 in DOM/1 will not be discussed here), the domestic sub-contract should take care to identify differences between the two forms of sub-contract.

Need for cooperation between quantity surveyor, contractor and sub-contractor

If the arrangements set out above are to operate smoothly cooperation between the quantity surveyor, the contractor and the sub-contractor is essential.

The sub-contractor would, for example, be well advised to find out from the contractor the time of the month when the latter proposes to make his valuations for payments on account so that he, the sub-contractor, can submit details of his *'valuation of work carried out'* in sufficient time to ensure that the contractor may include it with his own when arranging his monthly valuation meeting with the quantity surveyor. As these meetings are usually at the same time of each month it should only be necessary for him to make a single enquiry shortly after he commences work.

The sub-contractor should submit his statement in a format which follows as closely as possible the requirements of NSC/4, clauses 21.10.2 or 21.11.2.

The sub-contractor should also have in mind the requirements of NSC/4, clauses 21.10.2 and 21.11.2 when sending the contractor, architect or quantity surveyor 'all documents necessary for the purpose of the adjustment of the Sub-Contract Sum' [or] for the purpose of computing the Ascertained Final Sub-Contract Sum', as the case may be.

Chapter 16

Claims for direct loss and/or expense

claim: a demand for something as due, an assertion of a right to something (OED)

A claim does not automatically arise because a contractor/sub-contractor has sustained a loss nor is it calculated by deducting the sum paid by the client from the sum which represents the cost incurred by the contractor.

From what is said in Chapter 13 it will be apparent that a sub-contractor should not find it necessary to make a 'claim'. Where certain matters are affecting the regular progress of the sub-contract works he may make a 'written application' that he is 'likely to incur direct loss and/or expense', etc. It is then up to the architect to ascertain the loss, etc. or to instruct the quantity surveyor to do so (NSC/4 cl. 11). Where the work has been varied the quantity surveyor is responsible for valuing the variations (NSC/4, cls. 16 and 17).

What reason, then, has the sub-contractor to prepare a claim?

It must be remembered that the sub-contractor is obliged to submit details of his loss and/or expense to the architect or quantity surveyor if requested to do so (NSC/4, cl. 13.1.1.3) and most sub-contractors are aware that applications for reimbursement are not always recognized by architects, so that the sub-contractors find it necessary to demonstrate the extent of their loss. As far as variations are concerned, the sub-contractor has a right to refer any dispute to arbitration (NSC/4, art. 3), so the sub-contractor must be able to assess the value of his work for himself.

Having established a need for the sub-contractor to be able to demonstrate his loss and make valuations we may consider ways and means.

Preparing a claim

A prudent sub-contractor prepares the ground for a future claim when he compiles his estimate. He does this by recording the basis of his estimate, his calculations, etc. so that he is able to show the extent of the information which was available at the time of tender. He should be able to demonstrate the methods and plans which he intended to use, should his tender be successful. An estimate is only as good as the information available for its preparation. It may be said that the sum of a claim is the difference between cost of the works as known to the sub-contractor at time of tender and the cost of the works in the light of *all* the information.

Records are important to the sub-contractor throughout the life of all his projects.

The irony is that if his records are good he is less likely to find it necessary to submit a claim. Claims and disputes are usually in inverse ratio to the quality of the sub-contractor's records.

The steps in the preparation of a claim are essentially the same as those used for method or work study or, for that matter, any systematic form of examination, analysis, diagnosis, etc.

Selection of matters

The range of matters which may provide the subject of a claim will be largely determined by the 'list of matters' in NSC/4, clause 13 or by the contractor's default or omission. The sub-contractor should not be too selective when considering which items to include because there will almost certainly be a degree of negotiation with the contractor in due course and the sub-contractor may feel it prudent to have some 'face-savers' at that time.

Nonetheless, the sub-contractor should not work on the 'settle-for-half' principle – merely include some spare capacity. Experience with negotiations of this sort often reveals that the items which were included almost as make-weights are those which are most readily accepted by the contractor or architect.

The selection of items should emerge from the initial reading of the files which the person preparing the claim must undertake. There is frequently no clear line between individual items.

Sources of information include: estimate buildup, tender data, sub-contract documents, contract documents, specifications, bills of quantities, contract drawings, drawing schedules, working drawings, drawings with free-hand sketches by the architect or consultants used to illustrate instructions, architect's instructions and variation orders, contractor's directions, correspondence with all parties to the contract, minutes of meetings, records of telephone and other conversations, foreman's and site manager's reports or diaries, programmes, manpower planning histograms, estimates from and orders placed with suppliers and sub-sub-contractors, delivery vouchers, plant and tool schedules, labour and plant daily allocation sheets, Meteorological Office records, progress record photographs, 'special event' photographs, applications for payment, quantity surveyor's recommendations and architect's certificates, codes of practice, British Standard Specifications, reports by Royal Commissions, etc., etc.

The above are all prospective sources of information for claims. The list is only indicative of the wealth of information which may prove useful – in effect, anything and everything.

Recording the facts

This is usually a matter of scrap-booking cuttings from the sub-contractor's files. Original documents should be photocopied. A single document may contain material for several separate items so a system of coded slips may provide a method of arranging events by 'item' in chronological order.

Size A4 sheets guillotined into four narrow slips with holes punched in the top left-hand corners for treasury tags make it possible for events to be shuffled into order and for additional events to be inserted. The copies of the letter or site meeting minutes, as the case may be, can be pasted on the slips and referenced. If claims for several different projects are being prepared simultaneously the slips may be made from different-coloured papers; yellow for contract A, blue for contract B etc., or each slip may be marked A, B, etc. near the item no. to prevent confusion.

This may appear an unnecessarily elaborate precaution but a claim may be in course of preparation for several months and a number of people may be involved. The sub-contractor's credibility is undermined if he claims for a contract A event on contract B! If the cuttings are too large to fit on single slips they may be accommodated on continuation slips (if cuttings are three or four paragraphs long) or folded over on a single slip.

As the claim preparation develops it may be expedient to combine one or two items. It is a simple matter to combine slips but not so simple to split them. The object of recording the facts is to facilitate examination and the 'stories' which the records tell are more credible if they are derived from a variety of sources. They should be related to time-scales and bar-charts may be useful for this purpose.

The line between 'recording' and 'examining' is blurred. Obviously the recorder will examine the facts to some extent whilst making the record but as far as possible assembling the facts should precede the examination.

Examining the facts
This frequently produces surprises for the preparer of a claim who has been involved in a project for much of its life. The items which he/she had, as manager, considered to be primary causes of delay and disturbance of the regular progress of the sub-contract works appear under examination to be less significant on paper. The converse applies.

The examination should compare anticipation/expectation with actual achievement. The sub-contractor, or his estimator on his behalf, prepares an estimate which provides the basis of his tender from the data which is available at that time. If the circumstances, which the sub-contractor had reason to believe would exist, have changed since he submitted his tender he is entitled to consider the effect of the changes.

Specifications, drawings and programmes should be examined for variations. The dates of revision letters or drawings should be recorded and the actual changes (which should be noted in the legend on the drawing if the architect and consultants have done their jobs properly), should be studied. All drawings which have been revised should have been retained by the sub-contractor and must be compared with subsequent revisions and with the notes in the legend. The importance of a systematic approach to estimating, contractor administration, and procedures for storing data is now apparent.

The findings are best recorded on bar-charts because the time when an event

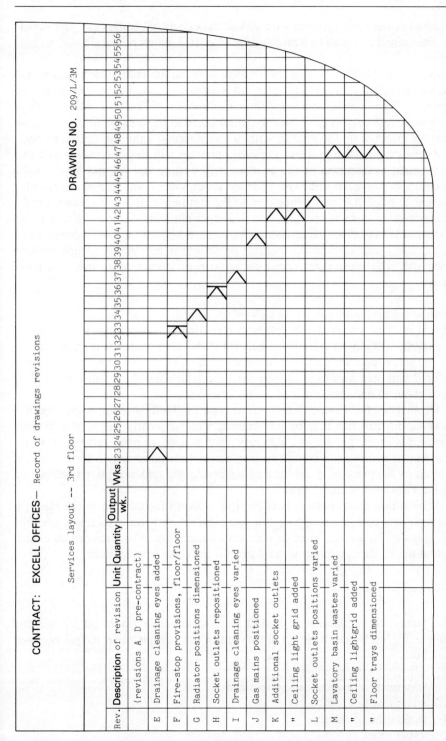

Fig. 16.1 Recording revisions of drawings

occurred is frequently a crucial factor and the effects of changes can more easily be determined if related to a programme. Fig. 16.1 shows how the bar-chart may appear.

In Chapter 13, Fig. 13.1 (p. 194) we saw how the sub-contractor's information requirements may be presented to the contractor and/or architect using the contractor's programme as a foundation. Comparison of Figs 16.1 and 13.1 makes it possible to ascertain (and later to demonstrate to the contractor and/or architect) the effect of information flow upon the progress of the sub-contract works.

As the examination proceeds the sub-contractor will almost certainly find that his original assessments of the situation will change and he may find that he simply does not have sufficient evidence about some items to make it possible for him to present a worthwhile case to the contractor. It is not enough for the sub-contractor to *know* he has suffered. It does not matter how sad the story his cost records may show; if his claim is to succeed he *must* be able to demonstrate that he has complied with the contract conditions in all the ways which we have discussed. The most experienced claims 'expert' cannot make bricks without straw.

Analysis (or diagnosis)
This is the task which should follow examination but in practice some measure of analysis will almost certainly occur whilst the examination is in progress.

The sub-contractor's analysis of the data which he has recorded, collected and collated will lead him to decide on the form which his claim will take or if he has a claim to make!

He analyses facts rather than fantasies and he should attempt to be objective however difficult this may be for someone who frequently has been intimately involved with the project and who may stand to lose (or gain) substantial sums. One way in which a sub-contractor may increase his objectivity is for him to put himself in the place of an arbitrator considering the evidence placed before him with a view to making an award.

If he considers an item to have merit it is worth including it in the claim; if not he should consider dropping it or, perhaps, strengthening it if this is possible. *The basis of his analysis must be the contract conditions.* The sub-contrator must have done all that he ought to have done if his claim is to succeed.

Format of claim

The format for a claim will depend to a considerable extent on its size and complexity but for the majority of sub-contractor's claims the principal headings may be:
(a)   sub-contract particulars;
(b)   statement of events leading to claim;
(c)   statement of claim;
(d)   evaluation of claim;
(e)   appendices.

Sub-contract particulars

These should identify:
(a)　the title of the project;
(b)　the parties to main contract and sub-contract (client, architect, main contractor, quantity surveyor, other consultants, and sub-contractor);
(c)　the tender;
(d)　specification or bills of quantities;
(e)　drawings, with details of numbers and revision letters;
(f)　articles of agreement (form of sub-contract NSC/4 with details of amendments, deletions, etc.).
Relevant dates and significant details should be given.

Statement of events leading to claim

This statement will probably open with a common theme indicating the manner in which the project progressed and the 'communication gap' widened. To facilitate cross-referencing the statement should be itemized. Dates and sources of information should be given.

The 'facts' recorded on the slips which were discussed earlier in this chapter are used for this purpose.

Entries in the statement of events might read:

2.18　　　Minutes of Site Meeting 9, (9.03) record that the service programme for
10 Feb. 80　the kitchen had been received and that a delay of three weeks had already occurred. 9.14 records that joist straps are still needed.
2.19　　　Letter . . . . enclosed drawing B/72/105 detailing holes through
13 Feb. 80　roof slab.

When the common theme has been established the sub-contractor is able to record the events related to individual items.

The statement of events should provide irrefutable evidence of the disruption (delay or whatever) which the sub-contractor has experienced so that a succinct 'statement of claim' may be prepared.

The entries should be 'factual' and capable of standing on their own. In some instances, however, they may require amplification by means of interpolated paragraphs.

Such paragraphs must be equally factual but they are important as a means of providing the 'story-line' and of interpreting the entries. They should also, therefore, be explanatory.

Typical paragraphs to follow the entries given as examples above might read along the following lines:

The constant flow of instructions referred to in items 2.10–2.24 above gave rise to varied and additional works valued at approximately £8,500. This sum represents a 30 per cent increase in the value of the work planned for the period November 1979 to the end of February 1980.

To appreciate the extent of the delay and disruption which was experienced each instruction should be appraised in the light of the amount of additional work required, the availability of labour and materials and the effect on the other work. In addition to the individual aspects, an examination of the relationship of each item with the others and their cumulative effect on the original programme sequence shows that neither sequence nor project performance could be maintained. Specifically:

(a)   there was a substantial increase in the volume of work to be executed;
(b)   deliveries of materials required for additional or varied work were extended;
(c)   labour resource requirements were subject to fluctuations.

Due to the volume of additional work and the uneconomical work resulting from disruption, work could not be executed in accordance with the programme and the planned labour resources were insufficient;

(d)   the enforced change of sequence and timing of operations resulted in disruption and uneconomic working;
(e)   it was necessary for operatives to re-enter areas of the building where work had previously been completed in order to undertake additional works.

The result of the above items was the need for constant planning, replanning and rescheduling of work and resources to accommodate the work arising from the numerous and frequent instructions and variations. The regular progress of the work could not be maintained and uneconomical working methods had to be instituted to reduce as far as possible delays which would otherwise occur. Productivity was impaired. The financial consequences of these delays, disruptions and actions are set out in the statement of claim.

Management involvement

The delays, variations and additional works, disruptions, etc. referred to above inevitably involve management in additional works. It may even have been necessary to employ an additional manager but at the very least the manager, engineer (or whatever he might be titled) responsible for the project in question will have found it necessary to devote additional time, which could otherwise have been spent on other projects, to sorting out the problems arising on the project with which this claim is concerned. The duration of his involvement will also have been longer in this instance.

Much of this involvement will be known to those concerned with the project and obvious from the entries in the statement of events but it must nevertheless be spelt out in the claim. Claims are frequently handled by separate departments when the client is a large organization and the sub-contractor must have in mind that his claim may, if all else fails, be dealt with by arbitration and an arbitrator can only arbitrate on what is put before him. Furthermore, the sub-contractor will probably wish to put a figure to his additional management costs and he must have provided an indication of the nature and extent of additional management involvement if he is to put forward a sum which carries conviction.

An introductory paragraph under a heading 'Additional Involvement of Management' might read:

The sub-contract works were scheduled for execution between . . . . During the first period the volume and nature of the works were such that the sums contained in the tender proved to be adequate. During the second period, however, considerable additional management was required to deal competently with the substantially increased volume of work and to reduce the delay and disruption caused as a direct result of variations to the planned progress of the sub-contract works.

Because of the number, nature, sequence and timing of instructions from the architect and directions from the contractor which affected the original works, additional management was fully engaged on the following tasks:

(a) attending meetings;
(b) preparing and amending schedules, enquiries and orders for sub-contract works, materials and plant;
(c) arranging cancellations and postponements in respect of the last;
(d) appraising and rearranging labour requirements;
(e) planning and replanning and supervising the additional and varied sub-contract works in conjunction with the main contractor;
(f) measuring and valuing variations;
(g) budget planning, accounting and financial control.

Similar details may be prepared in respect of other members of the management team.

Opinions vary regarding the format of claims generally but particularly with regard to the extent to which data is distributed between the statement of events and the statement of claim. On balance, however, it is desirable to extend the statement of events and to make the latter statement as brief as possible.

Statement of claim

A statement of claim may, then, read along the following lines:

From the foregoing statement of events it is apparent that:

(1) The contract period was from . . . to . . . a period of . . . weeks but the extended period was for . . . weeks, an extension of x weeks.
(2) The causes of disruption of the regular progress of the sub-contract works and the extended period were:

(a)	Variations (see summary of variations)	x weeks
(b)	Disruption (see Item . . .)	y weeks
		x & y weeks

The valuation of the additional costs is shown in the Summary of Additional Costs on p. 00.

Single-sum or detailed statement?
There is a school of thought which argues that as there are frequently so many and various interrelated causes and effects of delays and disruptions on building projects that no attempt should be made to attribute costs to individual items. It follows from this that a calculation of loss and expense is made, based on the difference between the cost actually incurred and the sum recommended by the contractor or quantity surveyor as the final account sum.

This 'single-sum' approach has the advantage of simplicity and the merit that it is more difficult for the architect, quantity surveyor or contractor to whittle away individual items.

The principal disadvantages of the single-sum are that it lacks credibility and it takes no account of the sub-contractor's inefficiency. It is difficult for a sub-contractor to convince a contractor or client that he (the contractor or client), should reimburse the sub-contractor in full. Similarly, whilst a sub-contractor may be convinced that he is 100 per cent efficient he should not assume that this opinion is automatically shared by all other parties to the contract who may argue: 'Why should we subsidize a sub-contractor's inefficiency?' But whichever approach the sub-contractor adopts, detailed statement or single-sum, calculations will have to be made.

No explanation is required of the single sum calculation which represents the difference between the sub-contractor's prime cost (with or without an addition for profit) and the sum which the contractor or client is prepared to pay in respect of 'final account'.

Calculations for a detailed statement of claim are another matter.

A claim is made for loss and/or expense incurred. The sub-contractor should therefore be able to establish the basis on which he prepared his tender so that he is able to prove that his actual costs have exceeded his estimate. The importance of preparing a tender in a logical, methodical manner is at no time more readily appreciated than when the sub-contractor is attempting to prepare a claim.

With the allowances in the tender for, say, management costs as the starting point and with his prime cost book as evidence of actual expenditure the sub-contractor is able to calculate the difference between tender and expenditure.

Management and supervisory costs

If, as was suggested above, an additional manager was fully employed on the project for a period of time the costs in connection with his employment are easily extracted from the prime cost ledger.

The additional cost of managers/engineers who would have spent part of their time managing the project in question but whom, by virtue of the delays, variations, etc. were more heavily involved, should be charged on a pro rata basis. The statement of events or the preamble to the statement of claim should indicate the tender allowance and the actual involvement:

'The tender allowance for the contracts engineer was one day per week and this proved sufficient during the original contract period. Subsequently in the extended period it became necessary for the contracts engineer to relinquish his duties on all other projects and devote the whole of his time to this project. The additional involvement was four days per week for the period from 1 July to 14 February, a period of 33 weeks.'

It is important that all such statements are numbered to facilitate reference to them in the calculations.

The cost of additional management may be calculated using the formula:

$$\frac{\text{Employment cost (£s per annum)}}{\text{Working weeks (or days) in year}} \times \frac{\text{additional time}}{\text{(weeks (or days))}}$$

Applying the formula to the 'contracts engineer' in the example given above and assuming:

employment cost   = £15,000 p.a.
working year   = 46 weeks
working week   = 5 days
additional days/week = 4 days
additional weeks   = 33 weeks

The calculation would be:

$$\frac{£15,000}{46 \text{ weeks} \times 5 \text{ days}} \times 4 \text{ days} \times 33 \text{ weeks} \ = £8,608.70$$

When assessing the 'employment cost' of the contracts engineer, care must be taken to ensure that *all* costs in connection with his employment are included in the claim.

The above approach may be applied to all staff, including head office staff, if it is not intended to charge the cost of their involvement as part of the 'establishment cost' by means of, say, a final percentage addition. The method of allocating establishment cost which was used when preparing the tender should provide the basis for the claim.

Supervisory visits

'Occasional' supervisory visits during the extended sub-contract period may be calculated in a similar manner to that described above or on a 'cost per visit' basis:

Charge for supervisor's time		
1 day per visit	60.00	
Travelling: 150 miles at 25p/mile	37.50	
	97.50	
35 visits at £97.50		£3,412.50

Labour costs

The labour costs of some specific 'extra' items of work may be clearly identifiable in which event they may be charged as such. In some instances, however, labour cost may have increased due to wage awards, non-productive overtime and travelling time (due to the need for men to work longer hours or travel from greater distances, etc. than was anticipated at the time of tender).

Such additional costs may be calculated for the extended contract period and listed in detail:

The following details have been extracted from time sheets submitted by operatives during the stated periods.

Labour increases from 1 July	Hours	Increase (£)	Total (£)
Chargehand	220.0	0.25	55.00
Technicians	150.0	0.20	30.00
Approved Electricians	320.0	0.17	54.40
Electricians	530.0	0.15	79.50
Labourers	466.0	0.12	55.92
Apprentices	245.0	0.10	24.50
from 7 August			
Chargehand	1,035.0	0.50	517.50
Technician	etc.	etc.	etc.
			£3,725.05
Fares increase	1,725 man/days	0.30	£ 517.50

Non-productive overtime may be calculated from time sheets making allowance for any provision for such time as had been included in the tender.

Loss of productivity may be calculated by assessing the ratio of labour:materials in the tender and comparing it with the ratio in the work as executed. If the tender was prepared by building up the tender sum by adding together the sums for labour, plant and materials, in a tender summary such as that shown in Fig. 4.4 (p. 56) the sums included are readily ascertained. In the summary in question it may be seen that the net cost of labour included in the tender is £18,867.40 and the cost of materials is £24,610.00. This gives a ratio of labour:materials = 43.40:56.60.

Assuming the prime cost ledger at the completion of contract records:

labour cost   £37,800.00
material cost   £31,600.00
a ratio of 54.47:45.53

and relating the higher labour ratio to the estimated labour cost included in the tender sum it may be seen that a loss figure *on the original work value* may be calculated as:

$$\frac{54.47}{43.40} \times £18,867.40 - £18,867.40 = £4,812.00$$

The calculated sum £4,812.00 relates to the *tender sum* and it does not take into account any variations which may have occurred during the course of the sub-contract works. These should be dealt with separately.

It will be appreciated that the ratios based on the prime cost ledger at the completion of the contract must be regarded as merely indicative. They do, however, provide a simple basis of calculation which, in the absence of a better method, enable the sub-contractor to demonstrate that he is at loss and that his records show a different picture from that which he anticipated at the time of tender.

Clearly, too, the ratios of labour: materials have changed from those on which his tender was calculated.

Labour is usually the most difficult ingredient of work to estimate, and to monitor and control during the course of the works.

It follows that the majority of a sub-contractor's losses are incurred through his failure to achieve levels of productivity which he considered to be reasonable at the time of his tender. This failure is frequently due to circumstances beyond his control.

Materials, plant and 'establishment' are, however, other causes of loss.

If labour may be described as the most *variable* ingredient it is true to describe material as the most *visible* ingredient – material is the product itself.

Assessing material costs

The material content of claims is frequently less subject to controversy than the other ingredients because of its visibility.

Comparison of tender with cost When assessing the extra cost of materials the starting points should be the sub-contractor's *tender* and his *prime cost ledger*. Comparison of the sums of these items will show loss or profit. If, to use the tender summary example shown in Fig. 4.4, the prime cost ledger sum does not exceed £24,610.00, the sum for materials in the tender summary, no action may be necessary in regard to the claim but if the prime cost sum exceeds the tender sum the sub-contractor must look further.

Materials abstract

An abstract of material contained in the prime cost ledger should be prepared which identifies the quantities of all items. This is most conveniently prepared on specially ruled abstract paper with sufficient space allowed on the sheet between items to make it possible for 'credits' to be entered for the materials for which the sub-contractor has been paid in the final account. The sub-contractor is paid for the materials in the final account under the headings of 'sub-contract sum' or 'variations' (see Ch. 12).

The quantities of materials to credit in the abstract against those contained in the prime cost ledger would, therefore, be taken from the sub-contractor's original estimate plus those contained in variations. Materials in variations which give rise to omissions are regarded as 'debits'.

An example of a part of an abstract is shown in Fig. 16.2 illustrates the method used. It will be seen that the prime cost ledger together with omissions contained in variation no. 17 account for 815 no. PK 79 (100 × 100) fittings but that the contract sum and the variations additions account for 730 units.

Quantity differences. What caused the difference between the number of units purchased and those used?

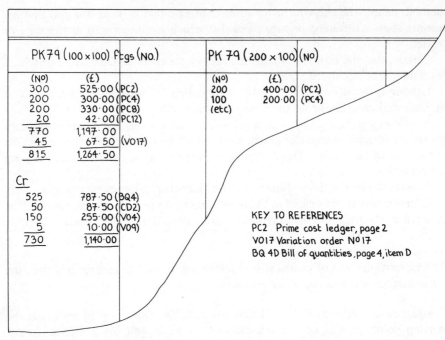

Fig. 16.2 Part of materials abstract

Possible causes are:

(a) estimating error. The estimator failed to measure the full extent of the work shown on the drawings. The sub-contractor will have to bear the loss (but profit from the experience);

(b) bills of quantities error. Similar to (a) but in this instance it is assumed that the sub-contractor tendered on the basis of a bill of quantities prepared by a quantity surveyor. The sub-contractor is entitled to payment in this event but he may have to measure the works as shown on the original drawings before the quantity surveyor will be convinced of the error/s;

(c) defective workmanship. Items of work incorrectly positioned, damaged during fixing, etc. will have to be made good at the sub-contractor's expense.

(d) theft or loss. Another contractual risk against which prevention is the best safeguard;

(e) additional works carried out in accordance with architect's instructions or with contractor's directions but which have not been accounted for as variations. The sub-contractor should check instructions against variations;

(f) waste due to breakages caused by the contractor or other sub-contractors after installation by the sub-contractor.

The important factor for the subcontractor is to identify the *difference*. It is then possible for him to ascertain the *causes* and submit a claim for items for which he may have a contractual entitlement.

Cost differences   In addition to variations in quantity, there may be variations in purchasing costs which will be identified in the abstract. Possible causes are:

(a)   inflationary variations since entering the contract;
(b)   variations in rates due to purchasing in small quantities;
(c)   suppliers' quotation errors;
(d)   invoicing errors.

Use NEDO formula

When there is provision in the contract for automatic cost adjustment by means of a schedule of basic rates or the NEDO formula, as occurs with 'fluctuation' contracts, the sub-contractor should have less difficulty in obtaining reimbursement than might be the case on 'firm price' contracts. On firm price contracts the sub-contractor may consider the use of the NEDO formula indices as a way of demonstrating the extent of his loss during an extended contract period. The effect of inflation, cause (a) above, frequently represents the major cause of a sub-contractor's loss. Increases resulting from small quantity purchasing rates, cause (b) above, should be recovered as variations if the materials were purchased in small quantities as a result of architect's instructions or contractor's directions.

Claims against suppliers

The sub-contractor may decide to claim against his suppliers for any loss which he may incur in respect of quotation errors and/or invoicing errors (cause (c) and (d) above).

Site supervision and management costs

Reference to the tender summary (Fig. 4.4) shows that site supervision and management were estimated £7,250.00. This sum would be derived from the sub-contractor's estimate of all the costs which he would incur in providing staff, plant, storage sheds, offices, telephone, etc., etc. in order to run the project. If he has used a check-list or standard pro forma for this purpose as shown in Table 4.1 (p. 51) it will be a simple matter for him to ascertain the cost of individual items which he included in his tender and thus calculate the additional cost which he has incurred as a result of the extended contract period, etc. He will also, of course, be able to justify his claim to the contractor by reference to his estimate, if he is called upon to do so.

Additional costs of providing offices, storage sheds, furniture, plant (such as benders, steps, ladders, extension leads, welding equipment, scaffold towers, etc.) may be assessed on a hire cost per week basis and the total cost per week may be multiplied by the number of additional weeks due to the extended contract period.

The matter of site staff, supervisors and managers was discussed above when methods of calculating additional costs were explained but the sub-contractor may prefer to calculate the *additional cost* of site management using the formula:

Site management cost × no. weeks in extension

No. weeks (or days) in original contract period

= £7,250 × (say) 10 weeks

           (say) 15 weeks

= £4,833

Quantity surveyors may be less inclined to accept such 'single sum' adjustments but in the absence of a detailed build-up of the sub-contractor's estimate there may be no alternative method of calculation. The sub-contractor should experiment with alternative methods of calculation to ascertain that which is to his best advantage. If, however, he decides on the single sum approach he should bear in mind that the quantity surveyor might well suggest that some items in his estimate would not be affected by an extension of time and disallow part of the sum. The cost of erecting and dismantling temporary offices and sheds are examples of items not affected by an extension of time. Whilst site management costs are normally time-based, being directly related to the duration of the project, some sub-contractors make provision for these costs as percentage additions.

Establishment (overhead) costs

The tender summary made separate provision for the addition of establishment (overhead) costs and for profit. The addition for establishment costs should take into account the share of running the firm which may reasonably be attributed to each project. This is frequently calculated as a percentage addition based on the ratio of estimated establishment cost to estimated turnover (sales). This is shown as 10 per cent in the tender summary and this percentage should be added to the total of the additional costs calculated above with the exception of site supervision if the format of the claim is to follow closely the format of the tender summary.

Financing cost

The contract makes provision for payments on account and the sub-contractor should not suffer loss as a result of the extension of the contract period, disruption, etc. The sub-contractor's claim should include a percentage, related to current interest rates, to reimburse him for:

(a)   financing costs on retention monies beyond the provisions in the contract. These costs may be calculated on the sum retained each month;

(b)   financing cost on outstanding balance during the period which it was outstanding, e.g.

$$\frac{\left(\begin{matrix} \text{sum of claimed} \\ \text{value of work} \end{matrix} - \begin{matrix} \text{sum paid} \\ \text{on account} \end{matrix}\right) \times \begin{matrix} \text{no. weeks sum} \\ \text{outstanding} \end{matrix} \times \text{interest rate}}{\text{No. weeks in year} \times 100}$$

Addition for profit   The tender summary shows an addition of 5 per cent on items 1–14 for profit. This percentage should be added to the calculated total.

Cost of preparing claim   This should be added to the total. These costs are more easily demonstrated if the claim has been prepared by a consultant who has charged a fee for his services.

Main contractor's discount   Nominated sub-contractors are required to allow the main contractor 2.5 per cent discounted and one-thirty-ninth should be added to the total.

The 'packaging' of claims is not an aspect which is taken very seriously in the building industry so the sub-contractor need not indulge in sophisticated artwork but he may find that photographs and illustrative diagrams are useful to make his case: 'A picture is worth a thousand words.' He should also provide detailed, supportive information in the form of appendices so that the story-line of his claim is not disrupted by unnecessary detail.

Presentation and follow-up

The sub-contractor may decide to deliver the claim personally – the personal touch has much to be said for it – and it may be expedient to follow-up by telephone after a suitable lapse of time. The 'first come first served' principle applies as much to claims as to other business matters. Following-up is also useful as a way of opening the way for negotiation. An enquiry regarding points on which the client may require clarification or may require further information can be used as the reason for the sub-contractor making the call.

Negotiation

Few claims are accepted in their entirety and a meeting between the parties may provide an occasion for negotiating on some items.

No two clients are the same so the level of acceptance of the claim will vary from client to client. The primary rule for negotiation is to avoid placing the other negotiator (and oneself) in an irreversible position. A first meeting might be planned with a view to sounding out the opposition rather than reaching firm agreement on cost matters.

The sub-contractor should, as suggested above, have provided his claim with some spare capacity. Whilst negotiation should not be conducted on the lines of an oriental market-place no negotiator would wish to report that he had spent time examining documents and attending meetings without, at least, obtaining a reduction in the claim in excess of the cost of the time he devoted to investigating it.

Acceptance

Unless the sub-contractor is unusually successful in his negotiations it is unlikely that his claim will be accepted in its entirety and he may have to decide whether or not to accept the sum which his client is prepared to pay.

At what point should the negotiator 'stick'?

Logic, even common sense, is seldom able to compete with emotion when it comes to negotiations and acceptance of a *'final'* offer but the economic fact of claims is that a balance must be struck between the offer which has been made to the sub-contractor, the possible award which an arbitrator (if the matter goes as far as arbitration) might make, the cash-flow embarrassment which the sub-contractor might experience whilst awaiting the arbitrator's award, and the cost of legal and arbitration fees if the sub-contractor's decision is to decline the final offer and refer the matter to arbitration.

As a ploy, a movement in direction of arbitration might lead to an improved *final* offer, but reference to arbitration by the sub-contractor is a serious decision which should not be made unless he is very sure of his grounds. Occasionally, as a matter of principle, arbitration may appear to be the only solution but principles can prove to be expensive luxuries.

Who prepares the claim?

The sub-contractor must decide who will prepare the claim. Will it be his own staff or a specialist in such matters?

The larger specialist contractor may well have an accounts or quantity surveying department which has personnel experienced and qualified in the claims field but smaller firms frequently do not have such departments and they must decide whether to arrange for their own staff to find time to take aboard the considerable work commitment involved in claim preparation or to engage a specialist consultant.

What are the advantages and disadvantages of the alternative courses of action?

These can be considered under several heading.

Available staff time

Unless the time of the staff who normally carry out estimating, surveying and the technical accounting functions within the sub-contractor's own firm is less than fully committed the staff may find it necessary to prepare the claim as an 'overtime' task. This may mean that processing the claim will be a prolonged task and that monies which the sub-contractor has outstanding will not be recouped in the near future. Furthermore, the sub-contractor may find a deterioration in his relationships with his client if the matter is not dealt with expeditiously. One of the banes of building contractors is the time which some sub-contractors take to submit their accounts. When considering sub-contractors for future projects, performance with regard to accounting matters will certainly be considered.

Staff knowledge and experience

The preparation of claims is not a task which, hopefully, a sub-contractor will be required to undertake on a regular basis and he will probably lack the necessary knowledge and experience for the task. The ingredients of a claim are very important and this chapter should assist the sub-contractor to be aware of the necessary 'input' but the recipients of a claim may take greater note of a submission which appears 'professional' than one which is home-spun. This will depend to a considerable extent on the client's degree of sympathy with the sub-contractor's cause. If the client has made it known that he is sympathetically disposed to a claim but that he needs dates, facts and figures before he would feel justified in recommending payment a sophisticated submission might be unnecessary and, indeed, even counter-productive.

If, on the other hand, the client is likely to resist a claim and the sub-contractor anticipates the need to resort to arbitration, a professional submission may be of considerable importance.

Cost of preparation

Cost of preparation is high regardless of whether the sub-contractor's own staff or a consultant undertake the work. If, however, the sub-contractor's staff are working at less than their potential capacity, in-house preparation may be more economical *provided* the expertise is comparable with that which should be obtained from a consultant.

A consultant will often give an estimate of his fee for making an initial examination of the papers and advising the sub-contractor regarding his case. Having ascertained the strength of the case, he may offer to prepare the claim, on a 'percentage of costs recovered', on 'a daily rate' or on a 'lump-sum fee' basis.

The 'percentage of costs recovered' basis is probably the most positive for the sub-contractor but consultants will not usually work on this basis unless they consider the sub-contractor's case to be good and the sums involved are sufficiently high.

A 'lump-sum fee' estimate ensures that the sub-contractor knows the extent of his commitment but consultants may pitch their estimates high because it is difficult for them to assess the extent of their work in advance.

Whilst most consultants will wish to take over the whole of the papers and do all the work within their own offices and employing their own staff a few will work with the sub-contractor's staff in an advisory capacity. In this event the sub-contractor undertakes much of the routine 'searching' with his own staff and assembles the information under the guidance of the consultant who is then able to interpret it and draft the various statements referred to earlier in this chapter. The largely routine clerical work, photocopying and assembly can usually be accommodated by the sub-contractor without undue disruption of his staff's regular work and at less cost than would be possible if the consultant used his own staff. Furthermore, the sub-contractor may obtain experience whilst working with the consultant which will be useful for the future.

The approach to preparing a claim which has been discussed above is not as detailed as that adopted by some consultants but it should provide the sub-contractor with an indication of many of items which he should take into account when seeking reimbursement.

Claims must be credible

It must be remembered that on very few occasions will the picture presented by the contract records be painted in just black and white. There will be shades of grey where the architect, quantity surveyor, contractor or, perhaps, arbitrator may have to exercise his discretion when considering the recommendation he will make to the client regarding the payment which should be made to the sub-contractor.

The human factor enters into the question of claims (it does in most aspects of management) and it is therefore important that the sub-contractor's claim appears to be credible. The sub-contractor should concentrate on clarity and simplicity of presentation. Architects, quantity surveyors and builders are sometimes suspicious of claims which appear to be 'clever' and to react against them. There is much to be said for the sub-contractor having the person to whom he is submitting the claim on his side.

Chapter 17

Management skills and responsibilities

Skills

In order to be able to undertake the tasks discussed in earlier chapters the sub-contractor must have basic skills. These can in part be acquired although it is often suggested that skills are largely inherent. In this chapter alternative approaches and factors are discussed.

The decision-taking process

Decision-taking in a business context is usually also *risk*-taking. Most business decisions involve risk and if the sub-contractor is not prepared to accept risk he should seek a different occupation.

The element of risk should, however, be minimized in so far as it is within the sub-contractor's scope to do so and with a systematic approach to decision-taking he is able to at least take into account the *known* factors.

Approach 1
(a)   identify the problem;
(b)   determine the alternative solutions; and
(c)   select the best alternative.

A more analytical approach is:

Approach 2
(a)   define the problem;
(b)   state objectives;
(c)   formulate hypothesis;
(d)   collect data;
(e)   classify, analyse and interpret against the hypothesis;
(f)   draw conclusions, generalize, re-state or develop new hypothesis.

A third approach, which is taken from standard method study principles is;:
(a)   select (the method/issue to be studied/decided);

(b) record (all the related activities/facts/data/opinions);
(c) examine (the material recorded in (b));
(d) define/decide (on the basis of (c));
(e) develop (new method/course of action);
(f) install (the method/action developed in (e));
(g) maintain (the installed (f)).

The initial letters of (a)–(g) above produce the acronym SREDDIM which has aided countless managers to adopt a systematic approach.

Obviously the level of importance of the decision to be taken and the time available for decision-taking will be important factors in selecting the approach to be used.

Decision-taking methodology is founded in question-asking and the British Institute of Management has prepared Check-list No. 19 for decision-making which poses six primary questions:

(a) have you identified the causes and extent of the problem?;
(b) what circumstances will affect your method of dealing with the problem?;
(c) have you identified your long- and short-term objectives?;
(d) have you collected and organized all relevant information?;
(e) have you explored possible alternative solutions?;
(f) What action is required to implement your decision?

Under each heading there is a number of secondary questions (thirty-eight in all) which include questions such as:

Q.1 *Why has the need for a decision arisen?*

Q.4 *Is it an outward sign of a more complex problem?*

Q.10 *What are the practical limitations of time and cost?*

Q.30 *Would a decision model of the whole problem area be useful?*

Q.38 *Have you established a system of follow-through and control?*

Apart from the numerical approaches such as mathematical (decision) models – which rely on fallible human input and are therefore prone to give precise answers to indifferent questions – approaches to decision-making are long established but none the less useful if they are actually used. So often, decisions are made on impulse.

Rudyard Kipling's six serving-men provide an invaluable check-list:

I keep six honest serving-men
  (They taught me all I knew);
Their names are What and Why and When
  And How and Where and Who

If the six interrogatives are used in conjunction with one of the three 'approaches' listed above they should assist with framing the right questions – an important part of decision-taking.

Major decisions

Major decisions in a business organization require the support of all concerned
for their implementation. There is, therefore, reason for consensus in the decision-
taking process. The phenomenal success of Japanese business organizations
has led to the examination of their decision-taking methods (Drucker 1979). The
Japanese approach is to devote time to defining the questions to be asked rather
than looking for answers. Whereas the Western approach has tended to start with
those concerned in decision-taking on opposing sides – with the losing side having
a vested interest in proving at a later date that they were right after all – the
Japanese avoid the 'my mind is made up, don't confuse me with the facts' stance
by attempting to define the *questions*. Much consultation is involved but when
the questions have finally been defined the answers are found quickly.

There is scope for dissent in defining the questions, so that all available
opinions and facts are examined, but confrontation which might lead to disruption at a later date, when the effects of the decision are known, are avoided.

Decisions based on opinions rather than facts

The argument is that many 'facts' are actually *opinions expressed as facts*. Indeed,
it is almost always possible to muster facts which fit the conclusions which have
already been drawn. Primary questions to ask which experimentally test opinions
against reality are:

(a)   what do we have to know to test the validity of this hypothesis?;
(b)   what would the facts have to be to make this opinion tenable?;
(c)   what is the measurement appropriate to the matter under discussion and to
       the decision to be reached?

To summarize the analytical approaches used for decision-taking it is difficult to
improve on Peter Drucker's statement: 'The most common source of mistakes
in management decisions is the emphasis on finding the right answer rather than
the right question.'

Emotional involvement   is a serious deterrent to good decision-taking. Objectivity
and emotion are incompatible. The letter of complaint from the contractor which
prompts the sub-contractor to make an immediate, impassioned response by telephone or letter should almost certainly be put aside at least until the next day.
Very few letters of that sort require a reply by return of post.

Managing people

An unattributed definition of management is 'getting things done through
people'. One definition of *manage* is: 'To operate upon, manipulate for a purpose'
(OED).

An appreciation of the *human factor* is necessary to manage people. It is one
of the skills which is to a considerable extent inherent but the *hints of managing*
(Hynd 1978) shown as Table 17.1 provide a guide to points which should be kept
in mind.

Table 17.1 Hints of managing

DO'S
Know the job.
Know the people.
Know the tools—financial, technical, managerial.
Show a sense of purpose.
Show a sense of determination.
Show a sense of urgency.
Show a sense of humour.
Be fair and honest.
Be understanding.
Be discreet.
Be demanding, but ready to participate.
Be practical.
Be punctual.
Be loyal.
Be thorough.

DONT'S
Be too impetuous.
Be too ready to blame others.
Be sarcastic to subordinates.
Discuss seniors with subordinates.
Discuss subordinates with subordinates.
Be a snob.
Bully.
Try to excuse oneself or be dishonest.

Delegating

Delegation is a sub-skill of managing people which is defined as 'the process a manager uses to assign a task or part of a task to a subordinate' (Drucker 1979).

A manager may *assign* a task but he cannot delegate (assign) his *responsibility* for the task. Delegation is not abdication. Delegation is an important skill if the manager is not to become too involved in detail. It is one way in which the manager is able to demonstrate his professional skill and increase his experience of the 'human factor'.

The benefits of delegation may be considered from the standpoint of:
(a) the organization in which the manager operates;
(b) his sub-ordinates;
(c) his own.

The Organization
Delegation assists the functioning of the organization by:
(a) creating a working environment in which people believe they are *performing and contributing*, not just carrying out tasks or duties;
(b) providing people with an opportunity to take decisions which they will have to be instrumental in expediting, in which they have a vested interest in

256

success and which will contribute to the success of the organization;

(c) ensuring that people are trained and available for promotion as the organ-
nization develops and expands. (If, when they are trained, people leave the
organization it is the rate of development of the organization which is at
fault, not the people);

(d) ensuring that tasks are being carried out economically because they are
being performed by junior (lower paid) rather than senior people.

Sub-ordinates

Delegation provides a subordinate with an opportunity to develop:

(a) his ability to make decisions, by giving him authority to take decisions
within his capabilities and within a monitored framework which will allow
the decisions to be effective but which will ensure that poor decisions will
not have too great an impact;

(b) his knowledge of his job, by enabling him to obtain experience;

(c) his skills, by giving him an opportunity to practise and develop them;

(d) his sense of responsibility. (Whilst the manager *retains responsibility* for his
subordinates' actions, even if he delegates some of his tasks and authority
to him, the exercise of authority by the subordinate should increase his sense
of responsibility);

(e) his sense of involvement in the organization, by allowing him to participate
in assisting to run it.

The manager should endeavour to create an environment in which delegation can
flourish but at the same time to provide monitoring and controlling mechanisms
into his delegation. There should be rewards (which should be greater than the
penalties) for the successful subordinate.

Delegation should be arranged so that the subordinate is able to produce
results rather than simply carry out tasks and duties.

If the manager delegates successfully the well-established management principle that decisions should be taken at as low a level in the organization as possible
will be implemented and the organization and the people in it should benefit.

The manager himself

Delegation assists the manager by:

(a) relieving him of some tasks which subordinates are able to do as well if not
better than he;

(b) reducing some of the *crisis management* which he would otherwise have to
carry out (see Ch. 7);

(c) providing him with time to do other things which are of greater importance;

(d) giving him an opportunity to demonstrate that he is capable of managing
a team.

There is a positive relationship between good management and good staff. An
inability to delegate is frequently an indication of insecurity and lack of
confidence on the part of the manager – the manager who is reluctant to delegate
being afraid that he is putting his job at risk to his subordinate by demonstrating

that he is not essential to the running of the organization. Managers who are over-authoritative are frequently reluctant to delegate and tend to keep things to themselves from fear that disclosure will undermine their authority.

In practice, the manager who is able to delegate is usually improving his promotion prospects rather than reducing them. If he has proved that he can manage that team, can he not do so again at a higher level?

Delegating is an important aspect of 'managing his own time' to which reference is made later in this chapter.

Communicating

Communication is 'the transfer of meaning' from one person to another. Managing people and an ability to communicate are probably the two most important skills for the manager.

Communication has four fundamentals (Drucker 1979):
(a) it is perception;
(b) expectation;
(c) it makes demands;
(d) it is different from information but interdependent with it.

Perception
Perception is 'the intuitive or direct recognition of a moral or aesthetic quality' (OED). Perception depends on experience and for successful communication to occur the communicator and the recipient should have shared experience or the communicator should have the ability to talk or write to the recipient in terms of the latter's experience – in his *own language*.

The strip cartoon in Fig. 17.1 illustrates the problems of communication which can exist in the building industry where differences in educational and social backgrounds between members of the building 'team'(?) all too frequently occur.

Experience Includes the cultural, social, educational and environmental backgrounds of the parties to the communication. The spoken word is only part of communication. Gestures and unconscious *body language,* (both forms of *non-verbal* communication), contribute to the total communication package. It follows that an ability to translate body language is a useful skill for the communicator and that an appreciation of the *reasons* for communication failures is the first step to preventing such failures. Figure 17.1 provides interesting examples of body language.

Expectation
Experience and experimentation have established that people see and hear what they *expect* to perceive. The human mind attempts to fit impressions and stimuli into its own framework of expectation. Furthermore, the mind resists attempts to *change its mind*.

Fig. 17.1 Problems of communication (from *Architects' Journal*, 19 April 1972, by Hellman)

To increase success in communication the communicator must know what the recipient expects to see and hear, or there must be a signal from the former that the message to be transmitted is *different* from that which the latter would expect to receive.

One of Wolf's Laws, (PD–OR), is: 'In briefings to busy people, summarize at the begining what you're *going to tell* them, then *tell* them, then summerize at the end what you *have told* them.' This advice is particularly sound if the communicator is transmitting a message which is likely to be outside the recipient's expectation.

Making demands

There must be *willingness* on the part of the recipient to receive the message. The message must be credible.

Information: communication

Communication is a *means* for transferring information and not an end in itself. There is usually no shortage of information but there is frequently difficulty in transferring its meaning.

It is ironical that with the massive increase in the availability of information and the higher level of sophistication of communication media, both of which have occurred in recent years, the problems of communication have increased rather than decreased.

Managing his own time

Few sub-contractors/managers analyse the way in which they spend their time and cost it for effectiveness yet his own time is one of his most valuable resources. This applies particularly to the proprietors of smaller businesses. The British Institute of Management Guidelines for the smaller business (no. 13) provides details of the approach outlined in this chapter.

Recording present activities
One study method is for the sub-contractor to record the way in which he spends his time for a representative number of days. A timesheet should be kept with columns for the time of day, duration of activities, description of activities and other persons involved.

The analysis might be by:
(a) activity;
(b) function;
(c) subject.

Activities might include reading, writing, dictating, telephoning, attending meetings, travelling, etc.

Functions might include budgets, labour relations, selling, estimating, planning, etc.

Subjects might include projects on which the sub-contractor is engaged. For each approach suggested above the percentage of time spent on each activity, function or subject is calculated.

The analysis should ascertain:
(a) the activities/functions/subjects on which too little or too much time is being spent;
(b) time spent on activities which do not contribute to the enterprise and which might be eliminated;
(c) tasks which might be delegated;
(d) activities which might be grouped more effectively.

Potential time-saving areas
Areas which should be considered are:
(a) meetings – to decide if some might be eliminated, held more/less frequently (see Ch. 9);
(b) mail – to decide which items might be handled by others, outgoing letters which might be standardized;

(c) information/reports to be considered – to decide if reports, minutes, etc. are essential for the running of the business;

(d) telephone – to decide nature and extent of use.

Identifying weaknesses is an important move towards improving one's personal performance but success depends to a great extent on one's willingness to accept the self-discipline which will almost certainly be necessary.

Responsibilities

The sub-contractor has a number of responsibilities. The best known are his responsibilities *to* his superior/s *for* tasks which he has to carry out and *for* subordinates who lie within his span of command.

He may, as stated above when discussing delegating, delegate (assign) a task but *he cannot escape responsibility* for it.

His other responsibilities include those he has to:

(a) his organization;

(b) others within his organization;

(c) others with whom he does business;

(d) the community (environment);

(e) himself.

His organization

To act honestly and loyally to carry out lawful policy, to respect the confidentiality of information regarding the company which he learns during the course of his employment, to disclose any personal interest which he may have which might conflict with the interests of his employer.

Others within his organization

To provide leadership and a working climate for others which enables them to develop their full potential, to provide training, to ensure that his subordinates perform their duties.

Others with whom he does business

To ensure that contracts are honoured in letter and spirit and that services and goods comply with specifications, etc. to meet the client's needs.

The community

To participate in and contribute to community activities, to take all steps to prevent damage to the environment, to conserve natural resources.

Himself

To perform to the best of his ability, in a humane manner and with integrity.

In effect, he has responsibilities to himself and all with whom he comes into contact. But his responsibilities as a business person are those which he has as an individual. They may be epitomized in the words of the medical practitioners' Hippocratic oath:

Above all, not knowingly to do harm.

References

(1)   Drucker, P. (1979) *Management*. Pan Books.
(2)   Hynd, J. (1978) *General Principles of Management in Construction*. CIOB, Ascot.

Index